BOUNDARIES AND STATE TERRITORY IN THE MIDDLE EAST AND NORTH AFRICA

MENAS Studies in Continuity and Change in the Middle East and North Africa.

BOUNDARIES AND STATE TERRITORY IN THE MIDDLE EAST AND NORTH AFRICA

Edited by

Gerald H. Blake and Richard N. Schofield

MIDDLE EAST & NORTH AFRICAN STUDIES Press Limited
Gallipoli House, The Cottons, Outwell, Wisbech,
Cambridgeshire, PE14 8TN, England.

Published by
MIDDLE EAST AND NORTH AFRICAN STUDIES Press Ltd.,
Gallipoli House, The Cottons, Outwell, Wisbech,
Cambridgeshire, PE14 8TN, England.

1987

British Library Cataloguing in Publication Data

Boundaries and state territory in the Middle East
and North Africa. - (MENAS studies in continuity &
change in the Middle East and North Africa).
1. Territory, National - Middle East - History
2. Middle East - Boundaries 3. Middle East - Politics
and government 4. Territory, National - Africa,
North - History 5. Africa, North - Boundaries
6. Africa, North - Politics and government
I. Blake, Gerald H. II. Schofield, Richard N.
320.1'2'0956 DS63.5

ISBN 0 906559 28 6 Paper
ISBN 0 906559 29 4 Cased

Printed by Whitstable Litho Ltd., Whitstable, Kent

CONTENTS

LIST OF MAPS AND FIGURES

Chapter 7

Chapter 8

LIST OF TABLES

PREFACE

With the exception of Chapter 8, the chapters included in this book are based on papers presented to the International Conference of the British Society for Middle Eastern Studies at the School of Oriental and African Studies in London in July 1986. The indefatigable conference convenor Dr. Tony Allan originally proposed a session on international boundaries, and without his drive and enthusiasm it would probably never have taken place. The session proved to be a lively one, and the papers created sufficient interest to encourage us to seek their publication. We are grateful to Menas Press Ltd. for taking up the suggestion positively and for helping to make the publication process a pleasure.

It seems hardly necessary to justify the publication of a volume on the theme of state boundary and territory in the Middle East and North Africa. For better or worse, the mosaic of states created in the relatively recent past have become as much part of the region's environment as its physical geography, or the cultural and religious legacy of Islam. The shape, size, and location of individual states are among the basic ingredients defining inter-state relations, and the state system is the framework within which all political and economic activity takes place. We hope that between us we have made a small contribution to understanding the origins and evolution of state and territory in the Middle East. Many people helped us in the task. We are deeply indebted to Mr. Roger Watts of Menas Press Ltd., for his enthusiasm, encouragement and patience and Miss Linda Brady for typing the manuscripts. We also acknowledge Methuen and Co., for kindly permitting the reproduction of one of their instructive diagrams in Chapter 7.

Gerald Blake & Richard Schofield
November 1986.

INTRODUCTION

Conference papers do not always make happy bedfellows when thrown together in a single volume such as this, but we believe the following chapters have a highly coherent theme. First, they trace the story of the transplanting of the European state idea to North Africa and the Middle East in the nineteenth and early twentieth centuries, largely by the colonial powers. At the end of this process a state system emerged based upon European concepts of territorial sovereignty, with state territory outlined by a network of land boundaries totalling some 21,000 miles (or 34,000 kilometres), **(1).** The idea of the territorial state was at odds with the region's precolonial tradition derived from Islamic practice in which communal relationships, not territorial control provided the basis of sovereignty. In Chapter 1 Bahgat Korany considers some of the contradictions arising in the Arab world as a result of the taking root of this European territorial system. While there is still a strong belief in Arab and Islamic unity as confirmed by opinion surveys quoted by Professor Bahgat Korany, the nation state is increasingly accepted as the norm in the Middle East. States are creating their own infrastructures, distinctive landscapes, and even demographic characteristics. One result is a deepening obsession with territorial and boundary questions, and a substitution of state nationalism for the "pan state" idea.

These themes are taken by George Joffé in Chapter 2 which considers the creation of colonial interstate boundaries, and boundary disputes in North Africa. He shows how the colonial legacy has obliged successor states to justify colonial administrative creations in terms of a nation that is assumed to occupy state territory, even when the nation concept was irrelevant to the pre-colonial situation. Although the Organisation of African Unity (O.A.U.) resolved in 1964 to leave colonial boundaries in Africa intact, this has not prevented a number of serious territorial disputes arising in North Africa. Many disputes associated with European boundary delimitation also persist in Southwest Asia, as Nurit Kliot shows in Chapter 3 on the Egypt-Israel boundary in Sinai. This chapter graphically illustrates some of the processes described in the first two chapters; when reading Chapter 3 it is also worth recalling Bahgat Korany's suggestion that the rise of a Jewish state on Arab territory heightened Middle Eastern consciousness about territorial issues.

The partitioning of Cyprus, which occurred after the Turkish invasion of 1974 is arguably the indirect legacy of two former imperial powers in the Middle East, the Ottoman and the British. Having been excluded because of British rule from the largescale population exchanges of the 1920s which took place between Greece and Turkey, the Greek and Turkish populations of Cyprus remained territorially intermixed until 1974, **(2)**. The Christian and Muslim populations have now become almost totally separated between the north and south of Cyprus. In Chapter 4 Carl Grundy-Warr considers the origins of the 1974 **de facto** partition and examines the present territorial arrangements in some detail. While the partition line has become a source of tension and resentment, and prevents any normal interaction between peoples, it is really only a microcosm of the effects that many colonial boundaries had elsewhere in the Middle East.

Chapters 5 and 6 introduce a second major theme, associated with functional and resource disputes over international boundaries. While disputes over boundary alignment and territorial sovereignty are becoming rarer as states reach formal agreements over their boundaries, functional disputes are becoming more frequent. Disputes over resources, especially oil and gas, are nothing new to the Middle East. State water disputes which Ewan Anderson discusses in Chapter 5 are relatively new, but they could greatly multiply as the demands for fresh water continues to rise. Problems arise because of river basins divided between states, groundwater aquifers crossed by international boundaries, and disputed ownership of wells and oases. Solutions may be found if goodwill exists between neighbouring states, but in the absence of political understanding, some water resource problems could become acute. In fact this has to be borne in mind in connection with most international disputes across, or concerned with, their common frontiers. Few disputes can be isolated from their broader political context. Boundary disputes are often a symptom of deteriorating political relations rather than the substance of the quarrel. Richard Schofield illustrates this in his study of the Iran-Iraq boundary in Chapter 6. Superficially, the Shatt al-Arab dispute was a functional dispute arising out of Iran's reluctance to recognise Iraqi jurisdiction over waters used by Iranian ships, particularly oil tankers. Richard Schofield argues that Iran's desire for a change of boundary in the river Shatt al-Arab was not based on navigational needs, but on ideas related to state sovereignty and ancient rivalries. When the two states went to war in 1979 the real causes thus went far deeper than where to fix an international river boundary.

Chapter 7 by Gerald Blake deals with the important process of offshore state boundary delimitation. The modern partitioning of the oceans of the Middle East and North Africa takes us back to Chapters 1 and 2 because the starting-point for maritime delimitation is the system of states and international land boundaries which were the concept and creation of Europe. Middle Eastern states are increasingly interested in their offshore domains, and state territoriality is moving inexorably seawards. Just as there were favoured and disavantaged states created when land boundaries were drawn up, there are today states which will inherit large and productive maritime zones, while others will benefit little from the new ocean regime. In Chapter 8 Stephen Langford discusses two processes by which state sovereignty is extended offshore, through straight baselines and historic bays. Although controversial, both devices have been adopted by North African states to ensure absolute sovereignty over certain coastal waters.

Chapter Notes

(1) A.D. Drysdale and G.H. Blake, **The Middle East and North Africa: a Political Geography.** New York: Oxford University Press, 1985, p.76.

(2) Brian W. Beeley, 'The Greek-Turkish boundary: conflict at the interface', **Transactions, Institute of British Geographers,** Vol. 3 No.3 (1978), pp.351-366.

I. INTERNATIONAL BOUNDARIES & ARAB STATE-FORMATION

Bahgat Korany

1.1 Introduction

If the 1980s brought the state back into political analysis (Skocpal et al. 1985; Kazancigil 1985) the systematic study of the Arab state is still nascent. To research the above-mentioned topic, taking the territorial definition of the Arab state, three notices were given to the computer of a major university library to suggest references: Arab world and inter-state conflict; inter-state conflict and the territorial state in the Arab world; and the territorial Arab state and minorities. The computer's answer: no books were available. Subsequent research disclosed the existence of two recent books on the Arab state and on minorities.

If we widen the field to embrace the structure and process of the Arab state at large, we are frustrated by the lack of a "best" or consensual approach to the subject. Thus a survey of the existiing literature on such an enlarged concept of the Arab state shows the domination ot two approaches: political psychology (**i.e.** the personalisation of the state) and religion (**i.e.** the role of Islam). Thus political psychology, emphasising the role of the leader or the 'big man' (Khadduri 1981; Vatikiotis 1978), led to a romanticised version of events and the perception of history as the product of powerful individuals. As for the emphasis on the role of Islam, it is a part of the well-entrenched orientalist (Said 1978; Hussein et al. 1982) tradition, and has become at present a flourishing industry after Iran's establishment of the Government of God (Benard and Khalilzad 1984). Despite their many differences, what these two major approaches have in common is to slight and overlook - if not to exclude altogether - the role of external factors and outside forces.

Yet if dependence conditioned the historical evolution of the international periphery - i.e. Arab and Third World countries -, is not the exclusion a priori of powerful external factors an anomaly in both method and substance? Consequently, rather than political psychology or orientalism this paper's approach emphasises political economy (Hollist and Tullis 1985; Staniland 1985; Strange 1985) to attract attention to the close interconnectedness between the vagaries of the world system and the **territorialness** (as distinct from the nation-ness) of the contemporary Arab state. Thus, in balancing the literature's "skewedness" in favour of psychological and orientalist approaches, it is hypothesised here that external factors predominated in the territorial definition of Arab states.

In order to develop this thesis, this paper is divided into two parts comprising four sections. Part one elaborates the thesis about the primacy of the external factors, discusses existing definitional problems in the use of the concept of the state, and specifies the definition used here. Part two analyses the collision between grass-roots political culture (based on a "pan" concept, whether Islamic or Arab) and the "foreign" political culture (based on the territorial concept). Section one of this part surveys some findings of recent empirical social science research (published up till now only in Arabic) to trace the presence of the "pan" idea in the Arab world. Section two addresses itself to the rise and globalisation of the phenomenon of the territorial state and its importation by the Arab world. To develop how the present demarcation of the Arab territorial state is indeed a phenomenon made in Europe (the then Centre of the emerging international system), three historical phases are traced: a) the rise and characteristics of the Westphalia system of 1648 which ended Europe's wars of religion and initiated an international system of sovereign states; b) dismemberment of the Ottoman Empire and its integration into the European system; and c) the rise of the Mandate system in its place and the resulting Arab territorial states.

The conclusion pulls the threads together to draw the lessons from historical data in order to throw light on the evolving political landscape in the Arab world. For instance, the creation of the state of Israel and the increasing emphasis on resources (**e.g** petrol, water) in regional politics increase the salience of inter-state territorial issues. The chapter ends by raising a nagging question: if the globalisation of the territorial norm explains the origin and contradictory existence of the Arab territorial state, is the 'outside' enough by itself to explain its present consolidation and continuing 'victory'?

*This paper is part of a chapter in the project "State, Nation and Integration in the Arab World", directed by the Instituto Affari Internazionale (Rome), to be published by Croom Helm in 1987.

1.2 Part One: Thesis and Definitional Problems

In relation to the political organisation of the community, **el-turath** or heritage emphasises two basic concepts. First the all-inclusive pan-islamism with its corollary of **umma.** However, the 19th century saw the decline of Dar El-Islam and in 1924 the Caliphate was formally abolished as a political institution. Pan-Arabism with its emphasis on "one Arab Nation from the (Atlantic) Ocean to the (Arab) Gulf" was a secular attempt to cope with European patterns of political organisation while salvaging some parts of **el-turath.** These two concepts of political organisation - as well as the attempts at their "reconciliability" (Al-Quamiyya al-Arabiyya in Theory and Practice 1980) - still dominate Arab political discourse.

Contemporary inter-Arab relations, however, have been dominated not by the realization of the Islamic **Umma** or the single Arab state, but rather by polemics over these two issues. In the meantime, (sub-national) territorial Arab states - Algerian, Syrian, Libyan, Kuwaiti or Sudanese - have acquired international legal personality and are getting consolidated. This is one of the most neglected aspects of the penetration of European political ideas into the Arab value system.

Why does this gap exist between discourse and reality? The thesis of this paper is that the gap between "said" and "done" is a reflection of a contradiction between indigenous grass-roots political culture (which is either "pan" or particularistic-ethnic) and the imported elite political culture, which emphasises the nation-state as the frame of reference. This nation-state concept is a relatively modern invention (less than four centuries old) and is a consequence of the rise of a European inter-state system since the Treaty of Westphalia in 1648. Through colonialism, the Mandate system in the Inter-War period, and the post-War political (but not ideational) decolonization, the European frame of reference got generalized. Consequently, nation-state formation in the Arab world is both hybrid and in a transition stage, two causes of its tension. Intra-state pressures (ethnic, religious or tribal, and Kurdish, Berber or Negroid) and inter-state territorial conflicts are manifestations of this tension. Tense and contradictory, the Arab territorial state is getting increasingly naturalized to become the order of the day.

The penetration of the nation-state concept in the Arab world is not dissimilar from what happened in other parts of the Third World:

> "The West's cruellest joke at the expense of Africa", remarks Ali Mazrui, "is - the construction of two contradictory prison-houses - one incorrigibly and rigidly **national** and the other irresistibly **trans-national.** One is the prison-house

of the sovereign state, a fortress of political and military sovereignty. The other is the prison-house of capitalism, compulsively transnational and constantly mocking the very principle of national sovereignty" (Mazrui 1984: 290).

In the Arab world, the hiatus between the triumph of imported ideas and the resistance of indigenous reality can be shown in this characterization of the Jordanian "state". Indeed, Jordan (previously known as the British mandated territory of Trans-Jordan):

> "...has the distinction of having the most artificial boundaries, the poorest endowment in natural resources and the least developed feeling of civic loyalty of any country in the Middle East ... Its desert dunes were unsuited for agriculture, and no minerals were discovered. Its coastal city of Aqaba ... was undeveloped as a port and its possession, moreover, disputed by neighbouring Saudi Arabia. All imports, therefore, had to come via the Palestinian ports of Haifa or Tel-Aviv (and since 1948 via Beirut and Damascus)" (Rustow, 171: 53-54).

This indeed raises the problem of the definition of the state, not only in the Arab world but in general. For even though the state is increasingly at the heart of political analysis, no standard definition has been canonised yet. As early as 1926 MacIver discerned seven uses of the concept of the state. Jessop (1982) has found six conceptions in the classical Marxist literature alone. This is why Clark and Dear (1984) could plausibly assert the existence of eighteen different theories of the state. But Titus is the most discouraging. In 1931, in an article in the **American Political Science Review,** he claims to have identified 145 separate definitions of the state. This is why after reviewing different approaches and definitions, the International Encyclopaedia of the Social Sciences (143-150) concluded: "at present ... conflict and parochialism (among approaches) overshadow harmony and the search for common denominators. As a result it is impossible to offer a unified definition of the state that would be satisfactory even to a majority of those seriously concerned with the problem".

Basic differences in the definition of the state do in fact reflect the fundamental divide among different schools of thought in political theory. Two such schools are the rights theory of politics versus the power theory of politics. As an example of the first, Hegel defined the state as "the actuality of the ethical idea". As an example of the second, Weber defined the state as the organisation which "monopolizes legitimate violence over a given territory". (A Dictionary of Political Thought 1982: 446-

447). Though the international dimension is not mentioned, an important definitional component - territory - is emphasized.

Consequently, in this paper a state is defined as the establishment of an organized political authority in a recognised territory. Territoriality is thus an integral part of the concept of the state. But an objection might be raised: haven't so-called stateless societies enjoyed a territory? Granted. De Lapradelle, writing on international law in 1928, seemed to agree and stated that "the striking feature of the concept of the boundary is its universal acceptance" (9). The attachment of stateless societies to place or locus, however, could be ad hoc and relates much less to bounded areas than to scattered features of the environment: **e.g.** sources of water, windfall foods or grazing land. This basic difference between "stateless" and "state-nation" concepts of territory and boundary constituted the essence of the conflict between the **Ikhwan** brotherhood and Al-Saud during contemporary Saudi Arabia's state-formation in the late 1920s. International recognition sanctioned only the latter.

This recognition, **i.e.** sovereignty (identification of exclusive legal control in a society), is another basic component in the definition of the modern nation-state. The 19th century jurist John Austin, one of the most respected authorities on the concept, linked his definition of the law and implicitly the concept of the state, to the concept of sovereignty. He held that positive law, on which formal government rested, could only exist in conjunction with a determinate locus of power and ultimate authority. Here then are linked the concepts of territory, sovereignty and state, or formal government as Austin used to call it (International Encyclopaedia of the Social Sciences, Vol. XV: 143-150). At this legal level in the definition of the state, we are at the heart of the importation of European ideas by the countries of the periphery.

1.3 Part Two: The Encounter Between Grass-roots and Foreign Cultures

i) - The 'Pan' Concept as the Basic Frame of Reference in Grass-Roots Political Culture

A survey of the constitutions of the Arab countries reveals an explicit insistence on the identification of the state as Islamic and/or Arab, and this does not change whether the state is 'moderate' or 'radical' in political ideology. Thus Saudi Arabia - which does not yet have a written constitution - has set up an 8-man committee to draw up a 200-article main "system of rule" based entirely of Islamic principles. In the meantime, the King rules in accordance with the **shari'a.** The People's Libyan Arab Jamahiriya, on the other hand, proclaimed through its General People's Congress in Sebha (2-28 March 1977) its commitment to achieving Arab unity - and that 'the Holy Koran is the social code' of the country. Even Egypt which pioneered secularisation and europeanisation in the late 18th and early 19th centuries and who has the dubious distinction of signing the first peace treaty with Israel has its constitution declaring that its "system (is) based on ... the country's historical heritage and the spirit of Islam. The Egyptian people are part of the Arab nation who work toward total Arab unity ... The Islamic code is the principal source of legislation".

Why this solemn insistence on the 'pan' idea in the face of divergent political behaviour in inter-Arab relations? One answer is to be found in the prevalence of the pan-idea at the grass-roots level and hence the attempt by the different regimes to capture this 'aspiration' in their official political texts. Recent empirical surveys of indigenous political culture - especially in its pan-Arab brand - confirm this inference.

The choice of these studies is dictated not by their perfection, but rather because they are pioneering applications of social science techniques (**e.g.** content analysis and survey research) to gauge Arab political culture both in recent history and at present. Thus despite usual methodological headaches over the application of such techniques in the "non-Western world", the results of this research are still useful as **indications** of the state of indigenous political culture as regards the question of unity (as a counterpart to the territorial state). The two research teams conducted their work in the late seventies and published their findings in 1980 (Yassin 1980; Ibrahim 1980).

Yassin and his team content-analysed some historical Arab writings for the evolution of the concept of unity through four historical periods: a) pre-World War 1: b) the inter-war period; c) the two decades 1945-1965; and d) the post-1967 period. Without going into details of the analyses two findings and their variations across time have to be emphasized for our present concerns.

The first is the relationship between the Arab nation and the **Umma,** or Arabism and Islam. Thus whereas early writings on Arab unity were either ambiguous concerning the relationship or tended to emphasize the religious dimension, later analyses (**e.g.** El-Husari) tended to be increasingly secular. They thus dealt with Islam not as a theocratic basis of the Arab nation but rather as its principal cultural component. Thus Amin El-Rihani, for instance, emphasised in 1938 that the Arabs antedated both Christianity and Islam and that Arabism comes first and above everything. A year later, Samy Shawkat - Director of Iraqi Education - went in the same direction to emphasise the pre-

Islamic civilisation of the Arabs. Ali Nasser was even more explicit; "religious atheism is relatively unimportant, whereas denial of Arabism is a crime".

The second variation in findings across time concerns the priorities emphasised in the different writings. Whereas the first two time periods had at the top of the list of priorities the 'rationale for unity' (32% and 27.9% respectively), in the last two time periods this prime place is occupied by "obstacles on the road to unity" (20% and 49% respectively). Does this increasing concentration of "obstacles to unity" reflect the writers' consciousness of the growing ascendency of the concept of the nation-state and raison d'etat as a frame of reference?

In fact, the second research project (Ibrahim 1980), surveying Arab political culture on the issue of unity, conducted its work at the ebb of Arab nationalist thinking and increasing crisis in the Arab system associated with Sadat's visit to Jerusalem, the Camp David Agreements and Egypt's conclusion of a separate peace treaty with Israel. For whereas Yassin traced the evolution of nationalist ideas **across time,** Ibrahim intended to measure their existence at a **specific point in time, i.e.** 1977-1979. Consequently, the research technique privileged was not content analysis but survey research. Thus the project regrouped about 200 Arab social scientists who surveyed and analysed a sample of 6,000 persons of different social categories in ten Arab countries. These countries contained two thirds of the whole population of the Arab world and embraced its different sub-regions from the Fertile Cresent to North Africa. What could the findings tell us?

Despite the antagonistic mass-media campaigns between Egypt and the rest of the Arab countries and the ideological confusion reigning in the area, 78.5% of those surveyed believed in the existence of an Arab entity, and no less than 77.9% believed that this Arab entity constitutes one nation (with fully 53% of this percentage believing that this nation is divided at present by artificial boundaries). What about doing away with these "artificial" boundaries and driving for statehood? Only 24.9% believed the disadvantages of such a project predominated over its advantages, whereas 33.4% believed that the advantages and disadvantages balance-sheet was equal and 41.9% believed the advantages predominated. This last finding is amplified and reconfirmed when those sampled were asked about the effects of unity on them as **individuals.** As many as 69.1% thought it would be beneficial (compared to 26.8% who thought it would be harmful), and 81.7% thought it would be even more beneficial for their children (compared to only 14.5% who thought it would be harmful). Arab unity, then, is the wave of the future, not only from an abstract political or symbolic point of view, but from a daily down-to-earth personal one. Why is it then not realized?

There are, of course, many obstacles: 48.5% of those

sampled believed that the most important obstacles were those imposed by external forces; 32.5% blamed some Arab leaders, 13.9% attributed the failure to lack of political consciousness among many Arabs, and 4.9% attributed it to the unwillingness of some Arab peoples. This explains, perhaps, why only a little more than a third (or 36.4%) of those asked desired to achieve unity in the short term (i.e. five years) whereas 40.8% aimed at its realization within 10 years and 22.8% envisaged a longer term: 20 years. More revealingly, only 21.5% thought that the objective should be one state with a central government; whereas the majority (57.3%) preferred rather a federation based on common foreign and military policies leaving internal affairs to the jurisdiction of the federated components.

It is perhaps the golden compromise between the "return to history" of "one Arab nation" and the daily reality of the existing **raison d'etat** ? And does the emphasis on external obstacles to unity reveal consciousness of the historical weight of the outside in shaping the present inter-state Arab system? This brings us to the impact of the global system on the rise of the Arab territorial state.

ii) The Arab Territorial State: A Phenomenon Made in Europe

A) The Globalization of the Inter-State Order

In a nutshell, the present inter-state system in the Arab world is part of a global process, of the expansion of international society (Bull and Watson 1984) or global rift (Stavrianos 1981). Though increasingly world-wide at present, this global inter-state order only dates back just over three centuries, more precisely to the 1648-Treaty of Westphalia which concluded the 30-year war in feudalized Europe and sanctioned the nation-state as the corner-stone of an emerging world order. This post-Westphalia inter-state order followed at least seven pre-modern world orders (Bozeman 1960; Holsti 1967; Northedge 1976) represented by the ancient empires of the Near and Middle East, the Chinese Empire, the Kingdoms of Ancient India, the Greek City-State System, the Roman and Byzantine Empires, the political and diplomatic system of Renaissance Italy.

There is a controversy in international historiography about the rise of the modern inter-state system. Some (Bozeman 1960) put it before the peace of Westphalia and others (Hinsely 1963) put it after. Without going into the details of this debate, we can say that even if the embryo of the inter-state order predated Westphalia, the new international order came of age with that treaty. This view is supported by the shift in focus of the normative apparatus at the basis of a political order: law.

Thus so-called international law (Falk et al. 1985) witnessed a shift after Westphalia: from a naturalistic to a positivistic theory of legal norms. For instance, writing at the beginning of the 17th century H.Grotius reveals that the rules of international law were identical in any Christian society in which the Bible and "right reason" (in that order of importance) were the true sources of right and wrong. But writing in the 18th century the Swiss lawyer Vattel, had an entirely different conception of international society and international law, one more akin to present-day international norms. For Vattel begins not with the single Christian republic of Europe, which is by now a part of history but with the national state:

> "Nations or states are political bodies", Vattel remarks "societies of men who have united together and combined their forces in order to preserve their mutual welfare and security. Such a society has its own affairs and interests; it deliberates and takes resolutions in common, and thus it becomes a moral person having an understanding and will peculiar to itself and susceptible at once of obligations and rights" (as in Northedge 1976: 61).

This is the embodiment of the norm of **raison d'etat** in its most explicit form at the international level. Consequent international lawyers only elaborated on this idea to emphasise the legal bases of the present inter-state system: state sovereignty, territoriality and equality.

To conclude then: the present inter-state system of Westphalia evolved through a lengthy process from about the 17th century to the mid-18th century. At the beginning of the 17th century "the modern system began to take shape when the rulers of Europe just started to think of their business as essentially material and earthly, rather than religious, or in other words as the conduct of the affairs of the secular state: all which was not God's domain was Caesar's domain". And by the mid-18th century "there was the acceptance of Europe, no longer as a single estate, the interests and welfare of which had to be defended against the infidels by sovereign and people alike, but as a collection of separate entities with interests unique to themselves. These entities were regarded as associated together by principles of law and morality which echoed the older unity of the whole, but which might from time to time have to give way to the even greater necessities of the sovereign state" (Northedge 1976: 64).

This intrinsically European system became universalized to the rest of the world. In the Arab region it did so in successive stages: the co-optation of the Ottoman Empire into the European balance of Power in the 19th century, then the fragmentation of this same empire and its replacemant by the Mandate system during and after the First World War, and a mandate system that shaped the present inter-state Arab order following the second World War.

B) Integration of the Ottoman Empire in the European System

Though from the 14th to the 19th centuries the Ottoman Empire occupied, controlled and administered between a quarter and a third of the European continent, the Empire was in but not of Europe. Indeed, the contact between the Ottomans and post-medieval Europe represented a collision between two opposed political cultures. For the

> "...post-medieval European idea of the State - a territorially defined entity apart from ruler or dynasty, organized in accordance with man-made rules - was alien to Muslim political theory. Ottoman theories of state and government derived from the Muslim concept that God is the source of all authority and law, that government exists to enable the community of true believers (Muslims) to fulfil its obligations to God. The community, not the state, constitutes the basic Muslim polity, transcending all boundaries" (Naff 1984: 143).

Thus the Muslim polity represented Dar El-Islam, whereas non-Muslim political entities represented Dar El-Harb, abode of war. Between the two there could be no equality of sovereignty as Europe's international law theorised, but a spectrum of relations ranging from **jihad** to provisional truce.

Such a pattern of relations could be practised so long as Istanbul was militarily powerful enough to back up its version of international society. But with the abating of Ottoman military prowess and the challenge it started facing on the Asian frontier, in the Balkans and in the north, the "ruling circle of the Empire had turned in on itself, yielding to conservative religious incluences often shading into fanaticism. Corruption and self-serving factions destroyed the system that had produced the great leaders of the past, while the Ottoman sense of moral superiority stifled curiosity about Europe, its societies, governments and religions". (Naff 1984: 148).

Thus the stream of Ottoman concessions started and multiplied, discreet and unconscious at first but increasingly basic and imposed. Even though Ottoman rulers harbored feelings of Muslim superiority over the "infidels", they were becoming increasingly integrated into Europe's state-system and hence recognized and accepted the principles of Europe's inter-state order: equality of sovereignty, reciprocity of relations, European diplomatic usages, points of European international law such as extraterritoriality and the "Law of Nations". The huge ideological barrier to alliance with European powers and the practice of rules of the game was thus demolished. These rules of the game were to turn against the very survival of the Empire itself:

> "...the bonds of tradition were permanently loosened, the old institutions were shaken to their foundations, and the idea of reform itself acquired an enexorable momentum within Ottoman governing circles. One of the most significant changes was a more favourable attitude towards European civilization by influential Ottomans...As European travellers in the Empire became more common, as the European communities burgeoned, and as, at the same times, the number of special Ottoman missions to the capitals of Europe increased, westward-looking Ottoman officials - small in number but influential - led the Empire, often unwittingly, into Europe's international system.
>
> Since the Ottoman Empire had no permanent diplomatic representation in Europe, Europe brought its system to the **Sublime Porte.** Resident European envoys in Istanbul insisted on instructing the **Porte** in the niceties of European protocol. The abler among them succeeded in making the Ottomans the instrument of their sovereign's foreign policy in eastern Europe" (Naff 1984: 155).

The year 1789 saw the seizure of the Crimea by Russia, the first piece of Islamic territory to be taken by a Christian power. Militarily incapable of stopping its dismemberment, the Empire under Selim III increasingly sought European allies, and started with Gustav III of Sweden (following his attack on Russia) in July 1789. But the Swedes, contrary to their Treaty with the Ottomans and without even referring to it, signed a unilateral peace with Russia one year later (July 1790). Highly disappointed, Selim III stated to his Vizir: "this is a harmful situation...Infidels are so unreliable". THe Sultan tried to cope with his ignorance of Europe's **realpolitik** by more integration in the system: the most significant feature concerning diplomacy being the establishment of Ottoman embassies in Paris, London, Vienna and Berlin.

The following example shows the magnitude of Ottoman assimilation of the basic norms of Europe's inter-state system at the expense of earlier Islamic rationale. When in 1799, the grand Vizir informed Selim III that a squadron of Russian warships had entered the Straits unannounced, the Sultan commented:

> "My vizir, the **reis effendi** ought to take this opportunity to

> remind the Russian interpreter in an amicable way of the international rules of conduct and of the reasons for the clauses in the Treaty (with Russia) governing this matter. It is contrary to the canons of international law that a war fleet should enter a foreign port without prior notification and without specifying the number of vessels". (as in Naff 1984: 159-160).

With the French expedition into Egypt and the rise of Napoleon, the **porte** effectively joined the continental alliance against France to become the first and only non-Western de-facto member of Europe's network of alliances.

But as mentioned earlier, assimilation was not limited to the Empire's external relations. For throughout the Empire's process of internal reforms to cope with the external challenge, European ideas, institutions and systems served as the modal frame of reference for change. Thus around the mid-19th century both the Empire's upper political and social levels as well as its bureaucracy and education were Europeanised in language, ideas and methods. The last barrier to Europe's swallowing of the Empire finally fell.

The rest of the story - from Mohamed Ali's challenge to the arrival of the mandate system - is well-known. It is this latter system that needs now to be tackled in order to complete the story of the globalisation of the European inter-state system and its implantation in the very heart of Islam: the Arab world.

C) The Mandate System

The story of the triangular relations between European powers, Ottoman Empire and the Arabs during and after the First World War need not be detailed here. Suffice it to say that such history-making events as the 1916-Sykes-Picot Agreement, the 1917-Balfour Declaration, and the 1920-Mandate System had two common features in relation to our subject: they presided over the dismemberment of the Ottoman Empire and planted the seeds of fragmentation in the once-deemed-possible Arab State.

The Sykes-Picot Agreement divided most Ottoman possessions into spheres of influence of Britain and France. The Balfour declaration continued the tradition of dismemberment and also made sure that the emerging territorial state in Palestine would not be Arab. As for the Mandate System it affirmed that the Arabs were not ready to have even mini-nation-states and that others (**i.e.** Britain and France) would decide on the timing and form of such a political organization. Article 22 of the Covenent of the League of Nations speaks of "peoples not yet able to stand by themselves", whose "well-being and development

... form a sacred trust of civilization", consequently, "tutelage of such peoples should be entrusted to advanced nations ... as Mandatories on behalf of the League". One small consolation for the Arabs concerned, they were designated as "A" mandate.

For the colonial settlement of 1919 in effect defined three classifications of those dominated: the "A" peoples of the Middle East (Palestine, Transjordan, Iraq, Syria and Lebanon) which were considered "civilized" but not yet ready for independence. The mandatories were to give administrative advice and assistance and - in the case of Iraq and Syria - to "facilitate the progressive development" of these territories "as independent States", constitutions were to be drawn up within three years so that these territories would be able "to stand alone". The "B" category designated the tribal peoples of tropical Africa who could require an indefinite number of years or decades of economic and political advancement under European tutelage. As for the last category, "C", it designated the "primitive peoples of the Pacific" and the "Hottentots" of South West Africa, who probably would remain European subjects at least for a period of centuries, if not for ever. (Louis 1984: 201). How different then is the Mandate formula from the old-fashioned straightforward colonial bond?

The question should not be regarded as rhetorical. For the traditional idealistic school of history - still influential in many quarters - considered the enactment of the Mandate system as the triumph of internationalism over nationalism, the victory of Wilsonian ideas and those of the 1917-Russian Revolution, notably of self-determination. Thus the Mandate system was affirmed as a mid-way stop toward a pre-fixed destination, and was closely associated with the catchphrase (**i.e.** self-determination) that the British and French accepted only with great circumspection. This linkage Mandate system-self-determination might indeed have been the intention of some. Was not a Permanent Mandate Commission constituted in Feb. 1921 as an auxiliary political agency of the League? Were not the governments of the mandated territories obliged to present annual reports to the Commission so that this latter advise the League Council whether or not the conditions of each mandate were being strictly observed? Were not representations from the inhabitants of the various territories heard, and governors and other high officials summoned for questioning? All this is true.

But this was the formal aspect of the Mandate system, its paper-work. Its reality was somewhere else: in the hands of those who held the mandate power. And for them the Mandate system was not very different from colonialism. What differed was the humanising veneer or the legal cover. Thus Lord Curzon - in explaining Britain's final decision to endorse the principle of self-determination - put it explicitly in Dec. 1918:

> "I am inclined to value the argument of self-determination because I believe that most of the people would determine in our favour ... if we cannot get out of our difficulties in any other way we ought to play self-determination for all it is worth wherever we are involved in difficulties with the French, the Arabs, or anybody else, and leave the case to be settled by that final argument knowing in the bottom of our hearts that we are more likely to benefit from it than anybody else" (Louis 1984: 205).

At the normative level, the inter-state European system did indeed benefit more than anybody else, for whatever interpretation of the Mandate system adopted, this system's end-result, as it concerns our topic at least, was not different from the colonial relationship. Whether we are talking about colonized Algeria, Mauritania, Tunisia, Aden, or mandated Lebanon or Syria, the result was the same: the institutionalization and consolidation of territorial states in the image of the European pattern. In a different context, Helms reached the same conclusion:

> "The present Middle Eastern borders and thus the whole of the modern Middle Eastern state system are products of this mandate period. From the perspective of the indigenous populations, there was little historical rationale in their formulation except in the most general sense. Many of the borders were linear, as if drawn with a ruler, determined by Europeans to further their own ends, however well-intentioned and idealistic, and to expedite mandate administration" (Helms 1981: 188).

Because of its 'alien' origins, the implanting of the inter-state system would face two pressures: internal strains within the polity at the time of the system's institutionalization (**e.g.** the conflict between Al-Ikhwan and Ibn Saud, Lebanon and generally the role of minorities in the mosaic society of the Fertile Crescent), and territorial disputes once the inter-state system was established (**e.g.** the several territorial conflicts in the Gulf area, the Western Sahara conflict).

1.4 Conclusion

So the Arab territorial state is becoming increasingly implanted and naturalized. It is not an indigenous phenomenon and yet it no longer seems a foreign import. It is thus a hybrid product. Though its form represented the primacy and globalisation of the modern European political culture at the basis of the Westphalia inter-state order, its content is increasingly nationalised. Despite this existing contradiction between form and content, acceptance of it as "normal" is growing. People get accustomed to its presence, it becomes the order of the day, the standard frame of reference. Did not the 1945-Charter of the Arab League sanction its presence? Moreover, Arab states whose territorial statehood has been questioned or threatened seek territorial strength by emphasising the sanctity of the principle of territoriality even in their constitutions. Thus Kuwait's 1962-Constitution stipulates that "no part of her territory may be relinquished", and Lebanon's Constitution affirms as one of its basic principles that "no part of the territory may be alienated or ceded".

Moreover, principal phenomena of the regional landscape do reinforce the trend of territorial primacy, **e.g.** whether we think of the rise of the state of Israel or of petrolism. Thus the plans for partition of mandated Palestine and the rise of a Jewish State of 'part of Arab territory' heightened consciousness about territorial issues. Similarly, the repeated Arab-Israeli wars, the continuing 'violation of international borders', and even the discussion of possible peace plans increasingly push territorial issues to the fore of regional politics. Even the conclusion of a formal peace treaty did not lessen the ascendancy of these issues as witnessed by the conflict between Israel and Egypt around the tiny area of Taba.

As for petrolism - as territorial contentions in the Gulf tend to show - it too has been associated with the saliency of territorial issues. Oil discoveries or their possibility have given value to erstwhile parts of barren desert and made them overnight an issue of high politics. Consequently, one can forecast that growing attention to other resources - river waters and seabed possibilities - will give importance to inter-state territorial issues. This is where the distinction high/low politics becomes increasingly dubious.

Thus a seemingly low politics issue like water resources exploitation is no longer separable from a high politics issue like the establishment of the state of Israel. If the experience of the past is any indicator the period 1951-1967 witnessed - in the area of the Jordan River System alone - eleven cease-fire violations, with the majority involving the U.N. Security Council (Naff & Matson 1984: 36-37). And in the 'arid' Middle East with its growing 'stragegic' value of water resources, the future does not augur well. Even with present consumption patterns, Jordan and Israel will be short of water long before 1995. Because of its military superiority, Israel has tended to seize by force the water resources of its neighbours and "it is known that 40% of Israel's current supply originates outside the pre-1967 borders of Israel (Naff & Matson 1984: 231). Such disputes over vital resources increases obsession with the territorial imperative all over the region.

But the problems of vital water allocation exist also among the Arab territorial states themselves. Thus Iraq and Syria witnessed a minor incident in 1975 because of the water of the Euphrates. The situation could have developed into a major conflict if not for the mediation of Saudi Arabia (and the tacit support of the USSR) which made Syria release more water to Iraq "from Syria's own allocation" as a one-time gesture of good will (Naff & Matson 1984: 232). The problems between Syria and Iraq around the Euphrates, however, are dwarfed by the possible problems around the Nile River basin, given the extent of that territory, its magnitude of annual water flow, and the size of Egypt's and Sudan's population.

On the other hand, and though it heightened obsession with the territorial imperative, regulation of strategic resources attracts attention to the inherent and glaring limitations of the territorial state, as the increasing pace toward integration even among established European states shows. And yet the Arab territorial state is not only resistant, but seems to be increasing in number. Are not the 'revolutionary' Palestinians and Sahraouis, who differ in (almost) everything else, united by their taking arms in the name of achieving territorial statehood? An important question should then be raised and explicitly: if the contradictions of the Arab territorial state are closely related to its foreign origin, does this foreign origin provide sufficient and necessary reasons to explain also its continuing consolidation and supremacy?

Though this is a new research problem that goes beyond the question privileged here, it is a very legitimate problematique (Ben Dor 1983). Suffice it to suggest for the moment that consolidation and supremacy of the territorial Arab state could also be one by default, a victory facilitated by the absence of an alternative, i.e. the failure up till now to elaborate an operational formula of the 'pan' state, Islamic or Arab-nationalist. Indeed, the neglect by Arab nationalist ideology to elaborate on the organisation of the desired Arab state is one of the most important loopholes in the nationalist doctrine. In this sense too, the failure of the short-lived United Arab Republic (1958-1961) has, in reality, only boosted the existing territorial Arab state. Would the Islamic fundamentalists fare better and come up with the needed alternative?

References

Bakhash, Sh. (1984), **The Reign of the Ayotollahs,** New York, Basic Book.

Ben-Dor, B. (1983), **State and Conflict in the Middle East,** New York, Praeger.

Benard, Ch. & Z. Khalilzad (1984), **The Government of God,** New York, Columbia University Press.

Bozeman, A. (1960), **Politics and Culture in International History,** Princeton, Princeton University Press.

Bull, H. & A. Watson (eds.) (1984), **The Expansion of International Society,** London, Oxford University Press.

Chabry, L. & A. (1984), **Politique et Minorités au Proche-Orient,** Paris, Edition Maisonneuve et Larose.

Enayat, H. (1982), **Modern Islamic Political Thought,** Austin, University of Texas Press.

Falk, R; F. Kratochwil; & S. Mendlovitz (eds.) (1985), **International Law: A Contemporary Perspective,** Boulder, Col., Westview Press.

Helms, Ch. (1981), **The Cohesion of Saudi Arabia,** Baltimore and London, The Johns Hopkins University Press.

Hill, N.L. (1976) **Claims and Territory in International Law and Relations,** New York, 1976.

Hinsely, F.H. (1963) **The Power and the Pursuit of Peace,** Cambridge, Cambridge University Press.

Hollist, L. & F.L. Tullis (eds.) (1985) **An International Political Economy,** Boulder, Col., Westview Press (International Political Economy Yearbook, Vol. 1).

Holsti, K. (1967) **International Politics: A Framework for Analysis,** Englewood Cliffs, Prentice Hall.

Hourani, A. (1962) **Arabic Thought in the Liberal Age 1798-1939,** London, Oxford University Press.

Hudson, M. (1977) **Arab Politics,** New Haven, Yale University Press.

Hudson, M. (1968) **The Precarious** Republic: **Political Modernization in Lebanon,** New York, Random House.

Hussain, A; R. Olson; & J. Qureshi (1984) **Orientalism, Islam, and Islamists,** Brattleboro, Amana Books.

Ibrahim, S.E. (1980) **Arab Public Attitudes Toward the Question of Unity,** Beirut, Center for Arab Unity Studies (in Arabic).

International Encyclopaedia of the Social Sciences (1968) (article by Morton Fried on the State: The Institution) vol. XV, p 143-150. New York, Macmillan and the Free Press.

Kazancigil, A. (1985) **L'Etat au Pluriel,** Paris, Economics & Unesco.

Keesings Reference Publications (1982) **Border and Territorial Disputes,** edited by Alan J. Day. Detroit, Gale Research Company.

Khadduri, M. (1981) **Arab Personalities in Politics,** Washington D.C., The Middle East Institute.

Korany, B. & A.E.H. Dessouki et al. (1984) **The Foreign Policies of Arab States,** Boulder, Col., Westview Press.

Korany, B. (1983) "The Take-Off of Third World Studies", **World Politics,** vol. XXXV, no. 3 (April): 465-487.

Lambton, A. (1981) **State and Government in Medieval Islam,** London, Oxford University Press.

LaPradelle (De), P. (1928) **De la Frontière: Etude de Droit International,** Paris.

Louis, W.R. (1984) "The Era of the Mandate System and the Non-European World" in Bull and Watson, **op. cit.,** 201-213.

Mazrui, A. (1984) "Africa Entrapped" in Bull & Watson, **op. cit.,** 289-308.

Nabarawi (el), F. & M. Mehanna (1984) **The Evolution of Political Thought in Islam,** Cairo, Dar el-Maarif, 2 volumes (in Arabic).

Naff, Th. (1984) "The Ottoman Empire and the European States System", in Bull and Watson, **op. cit.,** 143-170.

Naff, Th. & R. Matson (eds.) (1984) **Water in the Middle East,** Boulder, Col., Westview Press (in cooperation with the Middle East Research Institute, University of Pennsylvania).

Nasr, M. (1981) **The Arab Nationalist Conception in the Thinking of Gamal Abdel-Nasser 1952-1970,** Beirut, Center for Arab Unity

Studies (in Arabic).

Northedge, F.S. (1976) **The International Political System**, London, Faber.

Al-Quamiyya al-Arabiyya in Theory and Practice (1980) Beirut, Center for Arab Unity Studies (in Arabic).

Al-Quamiyya al-Arabiyya wa al-Islam (1981) Beirut, Center for Arab Unity Studies (in Arabic).

Rosenau, J. (1985) "The State in an Era of Cascading Politics", paper presented at the **International Political Science Association Congress**, Paris, July.

Rustow, D. (1971) **Middle Eastern Political Systems**, Englewood Cliffs, Prentice-Hall.

Said, E. (1978) **Orientalism**, New York, Vintage Books.

Scruton R., (ed), (1982), **Dictionary of Political Thought**, London, Macmillan Reference Books.

Skocpal, Th; P. Evans & D. Reuschmeyer (eds.) (1985) **Bringing the State Back In**, New York, Cambridge University Press.

Staniland, M. (1985) **What is Political Economy**? New Haven & London, Yale University Press.

Stavrianos, L.S. (1981) **Global Rift**, New York, William Morrow and Company.

Strange, S. (ed.) (1985) **Paths to International Political Economy**, London, George Allen & Unwin.

Studies in al-Quamiyya al-Arabiyya and Unity (1984) Beirut, Center for Contemporary Arab Studies (in Arabic).

Tibi, B. (1981) **Arab Nationalism**, London, Macmillan.

Vatikiotis, P.J. (1978) **Nasser and His Generation**, New York, St. Martin's Press.

Watt, W.M. (1968) **Islamic Political Thought**, Edinburgh, Edinburgh University Press.

Yassin, S. (1980) **Content Analysis of Arab Nationalist Thought**, Beirut, Center for Arab Unity Studies (in Arabic).

2. FRONTIERS IN NORTH AFRICA

George Joffé

2.1 Introduction

Most of the regional problems in North Africa trace their origins back to the process of frontier creation during the colonial period. This is certainly true of the two major conflicts that dominate the region - the Western Sahara conflict and the Libyan occupation of the Aozou Strip in Chad. However, it also applies to past tensions with Tunisia over the Jafara plain and the Ghadames tripoint; to tensions between Libya and Niger over Toummo oasis; to the recent crisis in relations between Libya and Algeria over the Ghat border region; to the 1963 'war of the sands' between Algeria and Morocco; to the issue of Moroccan recognition of the existence of an independent Mauritanian state; and to the Algerian problems with Niger and Mali until the border demarcation in 1983. Even the issue of Maghrebi unity - currently an issue of considerable social and political significance in North Africa - has suffered from the problem of ill-defined borders, with the result that there have been two different unity treaties in operation in the Maghreb which, in practice, only work to prevent unity - the Algerian inspired Treaty of Fraternity and Concord, to which Tunisia and Mauritania belong, and the Arab-African Union, created by Morocco and Libya but in theory open to any other state to join.

In reality, however, the border problems of North Africa arise from the inherent confusion between concepts of sovereignty imported by colonising powers and those inherent in Islamic political theory. This is compounded by European administrative practice and the development of a colonialist-nationalist dialogue which presumed on a political vocabulary based on European style concepts of territorial sovereignty and precisely defined territorial administrative delimitations. It is the dissonance between precolonial and colonial concepts of border which has been responsible for the regional problems that face North Africa today. This has been intensified by the development of modern nationalism in the North African successor states to the original colonial regimes, a nationalism also expressed in territorial terms and that has exacerbated border problems. The most acute consequence of these developments has been the conflict over regional hegemony between Morocco and Algeria, which currently finds its most acute expression in conflicts over border issues such as the Western Sahara problem and even the Libya-Chad dispute over which both regional powers have expressed interest and concern.

As is the case throughout the African continent and in large parts of the Middle East, the frontiers of modern North Africa are colonial creations. On superficial observation, there is little difference between the locations of the administrative divisions of the French, Italian and Spanish colonial empires in the region and the locations of frontiers today. In any case, the last word on such frontiers appears to have been said by the Organisation of African Unity (OAU) (OAU Cairo Assembly Resolution AHG/Res 17(1) July 17-21, 1964 and Article 3(3) of the OAU Charter) when the principle of the intangibility of colonial frontiers was established. The issue of principle having thus been resolved, only minor and technical problems can remain - or so it would appear.

In reality, however, there are still problems of jurisdiction over the Ghat region of the Algerian-Libyan border zone, disputes over sovereignty between Libya and Chad - to name but the most important territorial disputes - together with a mass of potential disagreements over maritime jurisdiction. These problems exist not merely because of inadequate delimitation and demarcation in the colonial period, but also because of the problem of nation building in independent North Africa and because of the ideological contradictions between colonial and precolonial concepts of the state that have persisted through the colonial experience and the advent of independence into a world of western style nation-states.

2.2 The Precolonial Situation

Nonetheless, the impression that the colonial period somehow resolved the major issues is even more strongly confirmed when the location of modern frontiers is contrasted with the precolonial situation. Although Morocco and Tunisia existed as independent or autonomous and virtually independent political entities (from 1837, the Beylik of Tunis accepted the formal suzereinty of the Ottoman empire, largely in an attempt to avoid European interference) frontiers - or more correctly **boundaries** (Prescott 1978; 31) in the modern sense of an alignment, a line allocating territory to states (Brownlie 1979; 3) - did not exist. Indeed, the only recognisable **territorial** distinction between states was a border march, a region of indeterminate political authority where military power provided the basis for political distinction between the authority of different states (Prescott 1978; 44). This status typified the border regions between the independent Hussainid Beylik of Tunisia and the Ottoman Vilayet of Tripoli (Joffe 1982; 19-42: Joffe 1984; 113-128) or the region between the Beyliks of Titteri and Oran in what is today western Algeria and the Moroccan sultanate.

Political entities corresponding to the modern states of Libya and Algeria simply did not exist. In the eighteenth century, the Regency of Tripoli had been a virtually independent state under the Qaramanlis, a dynasty of Ottoman origin. After 1835, Libya was composed of two Ottoman controlled **vilayets,** based at Tripoli and Benghazi, although Benghazi was really a **mutasarrifiyya** as far as military matters, justice and customs controls were concerned (Abun Nasr, 1975; 304). Outside the coastal strip, however, real power eventually devolved on the Sanusi religious order in Cyrenaica, particularly after the 1850s (Morsy 1984; 276) and on local tribes in Tripolitania and the Fezzan (Wright 1969; 102,112). After 1671, Algeria consisted of the Deylik or Algiers and the three Beyliks of Constantine, Titteri and Oran - all of which, although formally integrated under the Deylik of Algiers, enjoyed virtual autonomy as, indeed, did the Deylik of Algiers itself from its nominal suzerein, the Ottoman sultan in Istanbul (Abun-Nasr 1975; 175-176).

Further south, as the littoral regions of North Africa shaded off into the arid pre-Sahara zone and the desert itself, clear and continuous territorial divisions were completely lacking. As local populations became more and more concentrated around the available resources of water and land, so political control became correspondingly punctiform and, with increasing distance from the centres of political power along the coast, ineffective. Nomads paid little attention to formalistic considerations of political authority and sedentary oasis populations tended towards political autonomy, even if they recognised some overriding hegemony, based on religious or cultural associations.

Tuat oasis, now in modern Algeria, for example, recognised Moroccan hegemony because of a Moroccan conquest in the sixteenth century (Abun Nasr 1975; 213,297) and because of links with the Tihamiyya order in the north Moroccan town of Wazzan. There were even Moroccan officials there as late as 1891, as there were in In Salah (Morsy 1984; 210). The Sanusi order continually extended its area of control southwards from Jaghbub in eastern Cyrenaica towards northern Chad and westwards towards Tripolitania, while the Fezzan - at one time under its own autonomous sultanate - and Surt - originally controlled by the Awlad Sulaiman (Morsy 1984; 100-101) - accepted Ottoman suzereinty in the 1840s and extended Ottoman influence southwards towards Tibesti (Abun Nasr 1975; 304). By the end of the century, however, the dominant power in the Sahara region and in the Borkou-Tibesti region was Sanusi (Morsy 1984; 281-283).

The crucial feature of the precolonial situation was that political authority was expressed through communal links and was of varying intensity, depending on a series of factors involving, **inter alia**, tradition, geographic location and political relationships. The underlying consideration, however, was common throughout the region and involved a concept of political sovereignty that derived from Islamic practice. The essential consideration was that ruler and ruled were bound together through a conditional social contract in which the ruler could expect loyalty in return for ensuring the conditions in civil society for the correct practice of Islam (Zartman 1984; 151-152). Failure on his part to provide such conditions automatically authorised the community to replace him (Joffe 1986; 12-14: Mardin 1962: Lahbabi 1968).

It was this relationship that was encapsulated in the **bay^{c}ah** - which, in the North African context, came to be represented by a written document of allegiance and fealty from populations or groups within a population, to the ruler (amir, sultan, bey or dey) in which the restrictions on his authority were mentioned (Pennell 1986; 246-251). The crucial feature of this relationship was that it defined sovereign power and thus sovereignty itself in terms of **communal** relationships. It did not relate to **territorial** control, as the comparative European concept did (Thomas 1985; 3-5) but, instead, was primarily concerned with community and communal organisation.

There was, of course, a concept of territoriality, since populations occupied territory, but this was not the essential component, as the patchwork nature of political control in the Sahara made evident. Unpopulated territory could not, of itself, form the subject of sovereign identity (Zartman 1984; 152). Similarly, the confused situation of the border marches between Tripolitania and Tunisia arose because the limits of political authority were defined in terms of the recognised political links between local tribes and the authorities in Tunis or Tripoli. Territorial division had existed in pre-Islamic times - as with the Roman **limes** or the famous frontier between Cyrenaica and Tripolitania created by the Greeks and Carthaginians (Goodchild 1976; 17-21, 155-162, 195-196). However, the juridical basis that had underpinned them from the Classical period had been abandoned when the region was Islamised.

Furthermore, the conditionality of political authority in such peripheral regions was underlined by local assumptions of automonous tribal status (for the situation in Djerid region of Tunisia, see Hernia 1980; 188-190, 219-220: for the situation along the Tunisia-Tripolitanian border march, see Martel 1965;1, 401-410). The one exception to this practice, but one which only came to be practised after colonial ambitions in the region had already been made manifest in the wake of the French occupations of Algeria and Tunisia, was the Ottoman attempt to justify claims of territorial sovereignty extending into Chad (Lanne 1982; 30-31, 36-38).

2.3 Colonial Occupation

All this was to change with the advent of colonial occupation in North Africa. In the aftermath of the occupation of Algiers by the French army in 1830, a new system of administration was imported into the region. This was based on the concept of territorial demarcation of authority (Muller 1982; 165) - not just between different European colonial administrations, but between different types of administrative practice within the same colonial domain. The French occupation of Algiers, in short, marked a profound change in the nature of political authority - not just because alien authorities had replaced their indigeneous counterparts and had introduced a different political philosophy but also because the way in which that authority would be spatially applied had changed.

During the 1840s, French colonial control spread through the Beyliks that made up the equivalent of modern Algeria. In 1845, France was able to impose a border delimitation on Morocco in the Treaty of Lalla Marnia, after the Moroccan defeat at Isly in 1844. In 1881, after apparent provocation by the border Khumir tribes, France invaded and occupied Tunisia (Morsy 1984; 199) - on the grounds that its own control over the populations on its Algerian territories was threatened. In 1910, a border treaty was imposed on the Ottoman administration in Tripoli in order to delimit the territories of Tunisia and Tripolitania - less than a year before Italy invaded what was to become Libya.

In 1912, after several years of slow French penetration into Morocco from the east and from the coast around Anfa (later to become Casablanca) while Spain extended its influence from its **presidios** in Ceuta and Melilla which were occupied during the sixteenth century, France and Spain imposed a settlement on the sultanate of Morocco. This involved the problem of delimitation between the French and Spanish zones and was later to involve delimitations between different regions of French administration. It also brought to the fore the difficult problem of the future administration and frontier demarcation of the international zone around Tangier (Stuart 1931; 87-96), as well as reminding Spain of its ambitions for a Saharan empire.

At the same time, French colonial penetration into Chad and Niger from West Africa and southwards from Algeria was to link up the new colonial empire that had been created by France in North, West and Central Africa. The occupation of Chad began with the Battle of Kousseri in 1900 and was completed with the occupation of Tibesti in 1913 (Decalo 1977; iv-v). This immediately introduced the problem of the practical consequences of the 1899 Franco-British agreement over delimitation of their respective spheres of influence in central and east Africa and initiated the dispute over the border between Libya and Chad - a problem that began to receive attention as early as 1902. All in all, however, by the outbreak of the First World War, an initial delimitation of territorial frontiers throughout North Africa which was recognisably similar to that to be found after independence had been created.

2.4 The Creation of Colonial Boundaries

The modern boundaries of North Africa are a consequence of France's attempts to create an imperial domain stretching from the African coast of the Mediterranean into central Africa and then westwards to the Atlantic shores of West Africa. In the case of North Africa, therefore, they represent internal French administrative divisions as well as delimitations between France and its colonial partners - Italy and Spain. The principles on which such arrangements were made were, indeed, those already well-established by practice in Europe and, at the time, were uncontroversial. Details were often in dispute and were consequently frequently subject to violent disagreement, but the principles involved were not. Certainly, the details were to lead to disputes between independent states in the postcolonial period and then, of course, more profound disputes over questions of principle could - and often did - arise.

In the period before colonial occupation was complete, however, border delimitations were also forced on indigeneous authorities - the Ottoman administration in Tripoli in the case of the Tunisian frontier and the Moroccan sultanate for the western Algerian border close to the Mediterranean coast. Inevitably, the protracted negotiations involved, particularly in the case of Tunisia, a conflict of principle and ideological approach (viz Martel 1965; 1-67). Although France was able to impose its view, it did so at the cost of future problems for its successor states since the original indigeneous assumptions over the basis of division of sovereign political authority have persisted at some level throughout the colonial period and into independence.

A similar situation developed over border delimitations imposed by France - as the occupying colonial power - between territories that it considered an integral part of its own domain - Algeria - and those it considered protectorates where indigeneous authority was still theoretically sovereign despite effective French suzereinty. This was to apply to Morocco after 1912 (Brignon 1967; 341-356) and to Tunisia after 1881 (Nelson (ed) 1979; 33). Here too the innate conflict between indigeneous concepts of sovereignty and those imported through the colonial experience have continued to bedevil inter-state relations since the states of

North Africa achieved independence - 1951 for Libya, 1956 for Morocco and Tunisia and 1962 for Algeria.

The inherent juridical and ideological contradiction that was to mark all these negotiations was that, whereas Islamic political and legal theory and practice define sovereignty in communal terms - although it must be admitted that the provisions for this in Islamic legal practice (**shar**^{c}) are sketchy and inadequate (Schacht 1964; 76) - European jurisprudence - which has effectively become the basis for international law - places a primary emphasis on territory (Delupis 1974; 3-6). This is a consequence of the development of the European concept of the state as an area of unique and sovereign power over which political authority is exercised, rather than relating sovereignty primarily to populations (Joffe 1986; 7).

The ideological assumptions involved were inherent in colonial practice in Africa where little attention was paid to the reality of population settlement and migration patterns in establishing the divisions in territorial control, particularly when such divisions were established by metropolitan ministries and governments with no reference to the situation on the ground at all (Muller 1982; 166-167). On the other hand, great attention has always been paid to the exact location of the arbitrary lines that were defined as frontiers and boundaries of different political and administrative authorities , whatever the topological or demographic reality may have been - as would be expected from a **territorial** vision of sovereignty. Even during the colonial period - in North Africa at least - this approach caused a continual series of disputes and rebellions, over land ownership rights, migration patterns and pasturing traditions, thus demonstrating the consequences of too radical an imposition of a different set of ideological and cultural postulates on juridical distinctions of sovereignty.

2.5 Border location

The location of North African borders created during the colonial period is easily established from secondary sources. The most convenient and accessible general source for details of African borders and border disputes is Professor Brownlie's impressive study of African boundaries published in 1979 (Brownlie 1979) and much of the following discussion over the creation of North African colonial borders is taken from this source. Three other important secondary sources also exist, however - Martel's study of Tunisian and Algerian borders with Libya (Martel 1965); a general study of Algeria's borders by Boualem Bougaita (Bougaita 1979); and Trout's elaborate study of Morocco's frontiers (Trout 1969). Modern sovereignty and border problems are dealt with by Tony Hodges (Hodges 1984) for the Western Sahara and by Bernard

Lanne (Lanne 1984) for the Aozou Strip in northern Chad. The problem of the Algerian border with Libya in the Ghat region is not dealt with by secondary sources in any detailed way and here recourse must be made to archival material.

The following discussion will first outline border creation in North Africa, before turning to specific modern border and sovereignty problems. A final section will then discuss the theoretical problems involved in modern border structures and in concepts of political sovereignty. This will also involve a consideration of the degree to which European constructs have been absorbed into North African practice so that the congruence of the **nation** and the **state** has become of paramount importance in such issues.

2.6 Libya's borders

Libya's borders arise from three different categories of negotiations - negotiations over the location of existing precolonial boundaries (Tunisia-Tripolitania and Egypt/Sudan-Cyrenaica); negotiations over the interpretation of treaties drawn up between colonial powers or different administrations within a colony (the Ghat region and the Chad-Libya border); and negotiations over border locations in response to domestic or European issues (the 1935 proposed Chad-Libya boundary). Libya's eastern borders result from agreements between Egypt and Italy in 1925 and 1926, which superceded an earlier arrangement between the Ottoman empire and Egypt in 1841, and from agreements in 1934 between Britain and Egypt (as the condiminium powers in Sudan) on the one hand and Italy on the other. The latter agreement transferred the Sarra triangle to Libya - the territory allocated to Sudan under the 1899 convention and lying to the south of the 22°N parallel (Brownlie 1979; 102-109, 133-140).

The western borders - with Tunisia and Algeria - arise initially from negotiations between French authorities in Tunisia and Algeria and the Ottoman authorities in Tripoli. From 1902 onwards, these were paralled by discussions between France and Italy and these arrangements were later codified by further discussions in 1919. The final settlement was established in 1955 and 1956 by treaty between the newly independent government of Libya and the French government (Brownlie 1979; 27-43).

The border between Tunisia and Libya has been demarcated and is not in dispute, being confirmed by the Tunisian-Ottoman treaty of 1910 (Brownlie 1979; 141-148). The creation of the border itself, however, represents French success in delimiting the boundaries of border tribal territories and concerting this into a definitive territorial boundary by military pressure. It almost certainly represents an eastwards move from earlier conventions and is still the subject of confused land ownership patterns which transgress the border itself (Joffé 1984; 24, Joffe 1982; 34).

(a) The Ghat region

Although the border section between Ghadames and Ghat has not been open to dispute, having been delimited by negotiations in 1914 between France and Italy to Italian satisfaction - even though this represented a massive concession by France, compared with the putative border for the Fezzan and Algeria in the 1899 convention (Lanne 1982; 246) - the section to the south of Ghat is not so clearly acceptable, however. The only document laying down the principles of delimitation for the section between Ghat and the tripoint between Algeria, Niger and Libya (Point 1010) is the same as that which defined the overall delimitation down to Toummo oasis, the September 12, 1919 exchange of notes between Italy and France (Brownlie 1979; 127-132). This, however, merely lays down that the border should follow the line of mountain peaks, always leaving to Italy direct communication links between Ghat (at that time a vital transit point on the trans-Saharan trade route) and Toummo (also a significant location on the route which eventually extended to Ghadames and Tripoli, as well as travelling eastwards towards Egypt).

The problem is that this ambiguous phrase has been open to various interpretations. Although the Algerian interpretation suggests that the line passes through Anai pass, other interpretations exclude this, thus providing Libya with a large increase in territory towards In Ezzane. In part these confusions arise from the fact that Italy did not occupy the territory it had been allotted until the late 1920s and early 1930s and also because the more detailed definition of the border region was made as the result of internal French decisions. The organisation of the Algerian Sahara was a constant source of friction between the civil and military authorities in Algeria, the one constantly attempting to extend its control at the expense of the other. This was particularly important in the northern region of the Sahara, where civil authority was dominant. In the south, however, military power ruled supreme and here the major concern was to delimit Algerian spheres of authority from those in Niger and Mali - at that time both part of Afrique equatoriale francaise (AEF) and Afrique occidentale francaise (AOF) administrative complex.

A crucial conference was held at Bilma between representatives from Algeria, the AEF and AOF in 1932 to resolve administrative issues between the three French colonies. This, following on from an earlier conference held in Niamy in 1909, discussed the frontier issue with Libya and it is from the interpretations of the local topology made at this meeting that the conflicting border definitions now in vogue for the Ghat-Toummo section of Libya's western border derive. Even the reports of the conference are confused, since text and accompanying maps are in contradiction and the actual border line imposed seems to run in direct contradiction to the decisions of the conference itself. Subsequent meetings between representatives of the three colonies only further complicated the border delimitation issue. Although the existing arrangements for delimiting the border were confirmed by the 1955 treaty and the 1956 exchange of notes, no specific attention was paid to the confusion and thus the alignment of the actual border is still in dispute.

(b) The Aozou Strip

The southern border arises from a delimitation of zones of influence in Africa between Britain and France in 1899, but was later subject to further negotiation by Italy as the colonial power which culminated in the unratified 1935 treaty - the basis of the current Libyan claim to the Aozou Strip. The Anglo-French declaration of March 21, 1899 provided that the two zones of influence would be delimited by a line running south eastwards from the intersection of the 16th degree of longitude east of Greenwich and the Tropic of Cancer until it intersected the 24th degree of longitude. This arrangement was accepted by Italy in an exchange of notes with France in July 1902, although, until 1934, the southern border of Libya was completed to the east by the intersection of the 22nd parallel with the 1899 line (Brownlie 1979; 123).

Surprisingly, no map was included with the 1899 convention document - although the 1902 exchange of notes suggests there was such a map. If it did exist, it has never been found. This is important because the convention refers to a line running south east, whereas all later practice, treaty or otherwise - the 1919 Anglo-French agreement or the 1924 Sudan-Chad delimitation (Brownlie 1979; 618-639), for instance - make it clear that the tripoint for Sudan, Libya and Chad was to be 19°30'N, 24°E, indicating a line running east-south-east (Lanne 1982; 24-25). The tripoint would otherwise have been due west of Oum Chalouba in Chad. The western section of the border simply consisted of a line joining the terminal point of the Toummo oasis border from Ghat to the intersection of the Tropic of Cancer and the 16°N line of longitude.

The great problem with the Libya-Chad border arises from the interplay of European politics on the colonial African scene. This is the now famous Aozou Strip which Libya has occupied unilaterally since 1973 (Joffé 1981; 92), claiming that it is an integral part of its national territory on the basis of an unratified Franco-Italian treaty dating from 1935. Initially, the problem arose because, at the end of the Italo-Sanusi wars in Libya in 1927, the Fascist authorities in Tripoli began to lay claim to areas far to the south of the 1899 line, on the grounds that these had been controlled by the Ottoman administration of the **vilayat** of Tripoli and that Italy, as the successor state to the Ottoman administration in Libya, could claim the same territorial extent to its colony as had the Ottomans. The claim was implicitly based on the principle of **uti possidetis** - the legal principle that successor states are entitled to the international boundaries of their predecessors (Bougaita 1979; 233-274: Brownlie 1979; 11: Murty 1978; 169).

In a note to France on June 29, 1929, Italy laid claim to all Chadian territory north of the 18th parallel and to corresponding territories in Niger north of the same line and to the east of Ghat, along the 10th meridian east of Greenwich. The note was a riposte to a French proposal, made on December 28, 1928, for the cessation of the In Ezzane-Toummo-Djado triangle in Niger - a move clearly made to avert the growing demand in Italy for a major revision of the border situation in Chad (Lanne 1982; 108-121).

Although the Italian demand was rejected, France grew increasingly concerned over Italian behaviour in Europe, where the threat of a new war was closely bound up with the emergence of a commonality of interests between Fascist Italy and Nazi Germany as the 1930s wore on. At the same time, Italy had further demands on France over the status of Italian citizens in Tunisia (Lanne 1982; 125) as well as irredentist claims on Nice. In an attempt to counter all these problems and to ensure that Italy did not ally with Germany, the French government - despite considerable domestic opposition - decided to cede a large portion of Chadian territory, consisting of a strip of land parallel to the existing frontier and about 200 km deep to Libya. The proposed cession involved half of the Tibesti massif, the homeland of the Tebu - a tribal group which had close links with the Fezzan (Muller 1982; 170) and which had been closely allied with the Sanusi at the end of the nineteenth century (Morsy 1984; 282).

Ironically, the treaty never came into force - although it was approved by the parliaments in both countries, because the instruments of ratification were never exchanged. The reason was quite simple, shortly after the treaty had been negotiated, Italy joined the Axis and the occupation of Ethiopia became the major concern of Italian policy. In this climate, France and Italy had little to gain from the treaty and the border thus remained unchanged. However, the claim did not lapse and independent Libyan governments have continued to sustain it. Even though the 1955 Franco-Libyan treaty formally implied that Libya had abandoned any claims under the 1935 treaty, the fact is that the claim has been sustained, both under the Libyan monarchy and the Qadhafi regime. Indeed, a large part of Libyan policy in Chad has been conditioned by this consideration alone and Colonel Qadhafi was quite explicit in 1980 that this was the main aim of

Libya's intervention in Chad (Joffé 1981; 94).

The territorial division of the Tebu territory was hardly unique in the annals of North African colonial border delimitations, however, since the same problems had occured with the delimitation of the original border and no attention whatever had ever been paid to considerations such as pasturing patterns or cultivation locations of the different tribal populations involved. In this respect, the delimitation of the Chad-Libyan border represented the consequence of negotiations between sovereign states, rather than an imposition of European values of sovereign delimitation on local demographic divisions, as had been the case with the Tunisian-Tripolitanian boundary. In the latter case, problems had been caused by associating French concepts of territorial delimitation to indigeneous concepts of tribal autonomy whereby **usufruct** of the territory concerned was distributed between different groups, although political control was exercised **de facto** by the most powerful. The creation of an European-style impenetrable boundary had, in consequence, caused considerable hardship and stored up tensions and problems for independence. Although the frontier has now been accepted by both Libya and Tunisia, there are still periodic rumblings from Tripoli over the situation. The same conflict between demography and territorial delimitation obtained in the Ghat region, although there the issue of treaty interpretation is the primary concern today.

2.7 Algeria's borders

Algeria's borders with Libya have been discussed above and those with Morocco and the Western Sahara will be dealt with below. In this section, therefore, only the borders with Tunisia, Mauritania, Mali and Niger will be discussed. All of these states except Tunisia represent entities which, in their present form, are the result of colonial adminstrative delimitations - between the Algerian Territories du Sud (TDS) and the AEF for the eastern section of the Niger border from the modern Algeria-Niger-Libyan tripoint to the old Sarra triangle Sudan-Libya-Chad tripoint, and between the TDS and the AOF for the remainder of the Niger-Algeria border and those with Mauritania and Mali (Morsy 1984; 309).

(a) The Tunisian border

As far as Tunisia was concerned, the precolonial situation involved a delimitation between the Beylik of Constantine and the Beylik of Tunis. This, as was the case with the border march between Tunis and Tripolitania, was an area of indeterminate political authority which was linked to communal loyalties to separate centres of power in Tunis or Constantine. Insofar as a 'border' could be defined, this followed the delimitation created by the expansion of the Beylik of Tunis between 1609 and 1683 (Pitcher 1971; 143). This roughly followed the modern border line from the Mediterranean coast down to the border west of the Chott el-Djerid. It then swung eastwards, to parallel, first, the southern edge of the Chott before running into the Jabal Nafusa escarpment. The line was not impermeable, however, and the Tunisian Beylik considered that its authority extended towards the Saharan oases south of the Chott (viz Brown 1974; 132).

With the arrival of the French Protectorate in 1881, detailed delimitation and demarcation was undertaken between the Algerian administered areas of Constantine and Tunisia. This culminated in a definitive demarcation from Tabarka, on the Mediterranean coast, to Chott Rharsa by 1900. In 1901, an agreement between the Bey and the Resident General delimited the border between Chott Rharsa and Bir Romane, an agreement that was formalised on June 28, 1902 by the Bey (Brownlie 1979; 92).

The remaining section of the border was to show a sharp divergence from the precolonial situation, for a large element of the Sahara was included in Tunisia. This was largely the result of the fact that this delimitation involved the Algerian TDS administration and that it had to coincide with a delimitation undertaken with the Ottoman empire in Tripoli in order to establish a tripoint with Tripolitania close to Ghadames. According to the 1910 Ottoman-Tunisian agreement, the tripoint was to be located 15 km south of Ghadames, at boundary marker 223 and a subsequent demarcation in 1911 established this to be so. French administrative practice in Algeria, however, set the tripoint at boundary marker 220, to the north of Ghadames in 1911 and again in 1923. This delimitation was followed by metropolitan French practice after 1929 and, when a definitive delimitation andd demarcation of the border was agreed between Algeria and Tunisia after independence in 1970, it was this boundary marker that was chosen as the tripoint.

The Tunisian boundary with Algeria, therefore, represents a compromise between two different principles. The section to the north of Bir Romane has been delimited on the basis of a French interpretation of customary precolonial usage. The result was that the flexible and moveable boundary based on demographic and communal allegiance of the Ottoman period was replaced by an impermeable territorial demarcation which did not allow for traditional transhumant population movement and corresponded to European preconceptions of rigid territorially based administration. The section further south shows all the characteristics of a boundary created for administrative convenience in a region of sparse population. In fact, the failure of the border to correspond to population location in these regions led to constant problems with the Cha'amba nomads whose traditional movement patterns regularly transgressed it (Martel 1965; 1-109).

(b) The Saharan borders

The basic administration of the Algerian Sahara was laid down by a law passed by the French national assembly on December 24, 1902. This provided that all Algeria south of a line running parallel to the Mediterranean coast but about 200 km inland was placed under the direct control of the governor-general and would be subject to a separate administration. This was to consist of four separate military commands, each answerable directly to the governor-general (Journal officiel 1902; 8457-8458). In August 1905, the commands were defined as the territories of Ain Sefra, Oasis (Adrar), Ghardaia and Touggourt (AGGA 8H41). This decision came just after the ministries of the interior and the colonies in Paris had laid down the first basic delimitation between Algeria (TDS) and the AOF.

The delimitation had come about as a result of the French conquest of the South Saharan Touareg nomads (Trout 1969; 183) in 1904 and the conjunction of troops from Algeria moving south from the Tidikelt and troops from the AOF moving north from Bourroum. The line, as proposed, was designed in its eastern section to delimit Touareg territory so that it fell within the TDS. The western section was simply derived by linking up the line delimiting Touareg territory with Cap Noun, then assumed in Paris to be the southern limit of Moroccan territory. This was the result of pressure on the French government by local commanders in the AOF who wished to extend what was to become the administrative entity of Mauritania - which was to become a colony in 1920 - northwards while the Algerian administration wanted to occupy Morocco from the south (Trout 1969; 188).

The delimitation began south of the Ahaggar massif and north of the Adrar des Iforas in the Tanezrouft and followed physical features to east and west - rivers and valleys - until it reached In-Guezzam in the east and the mid-point of the route linking Marabouti and Taoudeni. To the west it then followed a geometric straight line towards the mid-point on the Fezzanese frontier between Ghat and the place where the route between Agades (in Niger) and Murzuk crossed the frontier (Brownlie 1979; 46).

The 1905 arrangements - typically based on administrative convenience and only in small measure concerned with delimitations linked to population location and movement in the Adrar - were in large measure confirmed by the 1909 Niamey Conference which was finally approved in 1911 by Paris. The modifications that were introduced concerned the southern and south western elements of the line and were solely designed to improve administrative control of the Touareg (Trout 1969; 191-192). Otherwise, the line remained unchanged, although all mention of Cap Noun as the southernmost extent of Moroccan sovereignty had been dropped - partly because, no doubt, of the arrangements that had been made with Spain over its ambitions in the Western Sahara.

The sole remaining problem was the accurate location of the two tripoints on the delimitation line. The tripoint between Mauritania, Mali and Algeria was finally defined by the Kayes Treaty, drawn up on February 16, 1963, which provided the definitive delimitation between Mauritania and Mali. Although Algeria was not a signatory to this treaty, there is no evidence that the Algerian government has ever objected to the location of the tripoint under the treaty, particularly as it corresponded to the location defined by the Niamey Conference and the restatement of its location under the 1944 government decree delimiting the French Sudan (Mali) from Mauritania (Brownlie 1979; 407-410). The location of the tripoint between Mali, Niger and Algeria is far more problematic since the only agreements that delimit Niger and Mali were French administrative measures and these are not exact. As a result, the tripoint has become accepted by international usage and Algeria has not disputed this decision (Brownlie 1979; 417).

Since the end of the colonial period - Mali, Mauritania and Niger all obtained independence from France in 1960 under the provisions of the Gaullist Communaute francais legislation, while Algeria obtained independence under the Evian agreement in 1962 - there has been no evidence of dispute over the Niamey delimitation. Demarcation, however, is another matter and there has been constant friction over precisely where on the ground the boundary line should run. On January 31, 1970, Algeria and Mali announced that they were to demarcate their common border, while a similar announcement was made by Algeria and Niger on July 28, 1971.

Mauritania and Algeria made no comment on their common border until 1983, when they announced that delimitation and demarcation had been achieved as part of the process of Mauritania's adhesion to the Algerian inspired Treaty for Concord and Fraternity. Similar announcements were made in conjunction with Niger and Mali in the same year. In this case, it is clear that the 1964 OAU Cairo Resolution provisions have been applied.

2.8 Morocco's borders

The definitive study of Morocco's borders with its neighbours was produced in 1969 by Frank Trout in his work, **Morocco's Saharan Frontiers.** He argues convincingly that Morocco was very much the victim of French administrative intrigue. The result has been that the boundaries of modern independent Morocco bear little resemblence to those associated with the precolonial political entity of the Moroccan sultanate. Trout specifically argues that the European powers involved in signing the Act of Algeceiras - the concluding document of the Algeceiras conference in which European powers, in return for guaranteeing the integrity of the Sharifian empire, extracted commercial benefits that were to contribute significantly to colonial occupation (Brignon 1967; 300) - failed to guarantee Morocco's territorial integrity; that France and Spain made a series of illegal agreements over Spain's occupation of the Western Sahara and that France allowed the Algerian Government-General to illegitimately remove Moroccan territory from the control of the Protectorate (Trout 1969; 432-441). The result of this has been that Morocco's post-independence frontiers have been amongst the most problematic in North Africa.

The situation has been complicated by the fact that the Moroccan nationalist movement, **Istiqlal** - under its charismatic leader, Allal al-Fassi was particularly active in making specific claims on post colonial border settlement in the 1950s. These involved claims to a hinterland south of Morocco and extending right down to the Niger river, encompassing the Western Sahara, Mauritania, Western Algeria and Western Mali (Hodges 1983; 86). This led to a prolonged dispute with Mauritania over the latter's right to independent existence and to a short border war with Algeria in 1963.

However, the major problem arising from the ambiguous colonial experience has been a dispute far more concerned with sovereignty than with border delimitation - the Western Sahara issue, where, since 1975, Morocco has occupied the region despite opposition from Algeria and an indigeneous national liberation movement, the Polisario Front. The affair has now acquired international ramifications, through the involvement of the International Court of Justice, the OAU and the UN, and has called into question the basic principles involved in the decolonisation process that have been developed by both the UN and the OAU over the past forty years. The issue involves, therefore, all the factors that have already been discussed earlier and demonstrates their clear relevance to current political situations, as well as their academic interest. It also brings the issue of UN decolonisation practice involving self determination into the problem, contrasting the validity of this approach with that of **uti possidetis** (viz Joffé 1986).

(a) Borders with Algeria

Morocco's borders with Algeria are the result of a complex series of interlinked problems of delimitation that relate largely to internal French administrative disputes. They also relate to the pattern of French occupation of Morocco, as French forces nibbled away at the periphery of the Sharifian empire in the closing years of the nineteenth century (Brignon 1967; 338). This was particularly significant in the period 1899-1901, although French encroachment continued up to 1910 before France was able to extract an agreement from the Moroccan sultan, Moulay Abd al-Hafidh, that further encroachment was in support of Moroccan authority (Dunn 1977; 26).

i) The Treaty of Lalla Marnia

The initial contact over boundary delimitation occurred in the 1840s, however, after the Morocco defeat by French forces at Isly in 1844. France was then able to require Morocco to sign the Treaty of Lalla Marnia in 1845, which delimited the border between the coast by Selwan and Teniet el-Sassi. This effectively codified the traditional division between the Moroccan sultanate and the Beylik of Oran - set up first after the Saadian repulse of Ottoman attempts to invade Morocco (Yahya 1981; 174-180). It also outlined a delimitation down as far as Figuig in terms of the allegiance of local populations and rejected the need for a delimitation further south because it was uninhabited, a good indication of the flexible and communal nature of local perceptions of sovereignty (see clause 6 of the treaty - Brownlie 1979; 62). The 1845 treaty was confirmed by protocols signed between Morocco and France in 1901 and 1902, which also made it clear that Figuig and Ich were to be considered Moroccan.

ii) The Varnier line

The situation was to be further codified in 1912, when Maurice Varnier was appointed High Commissioner for Eastern Morocco, as part of the process of French occupation. His delimitation was designed to provide a boundary between the administrative responsibilities of the Ain Sefra military command and those of the new French administration in Eastern Morocco. The portion south of Teniet el-Sassi more or less followed the 1845 treaty outline and the 1901 and 1902 delimitations. At Figuig, however, it then turned west and, following the mountain range along Jabal Grouz to the Oued Guir where the latter joined the Oued Zelmou. This line codified what had become French administrative practice in the wake of a further agreement between France and Morocco in 1910, whereby France recognised that Bouanane and Bodenib were Moroccan. French forces, however moved into the confluence of the Oued Guir and the Oued Zousfana, thus effectively treating these regions as part of Algeria. The westward projection of the line was not defined, except to warn that the Tafilalt and the basin of the Guir would have to be Moroccan (Trout 1969; 105-109).

The significance of this line was that it became the administrative boundary during the French Protectorate in Morocco, although the Moroccan authorities did not accept it, on the grounds that French forces had strayed into Moroccan territory before the Protectorate was declared in March 1912 and that the line merely codified this usurpation of Moroccan territory. In fact, although Varnier made many concessions to Algerian administrative demand, he did resist the extreme views of the French authorities in Algiers who wished to set the eastern Moroccan border at the Moulouya river. This stand was eventually confirmed by Marshal Lyautey once he became resident general of Morocco in 1912 (Trout 1969; 115). The result was that French Algerian forces were forced to abandon Figuig which they had occupied in June 1903 (Dunn 1977; 169).

The line itself had been based on a compromise between French treaty obligations to respect population centres clearly under Moroccan authority and French army attempts to extend its administrative authority as far eastwards as possible before the Protectorate was established. However, it did confirm another reality - the French occupation and alienation to its new administrative colonial construct of Algeria of far more distant populations and territories to which Morocco laid claim and where Moroccan representatives had regularly been appointed. Chief amongst these was Touat, which fell to French forces in 1901, along with Gouara and Tidikelt (Dunn 1977; 146). It is clear, however, that all these areas were considered by the Moroccan sultanate to be part of Morocco's overall sovereign sphere and Moulay Hassan I made considerable efforts to assert his authority over the Sahara that Morocco considered to be part of its own communal sphere. The problem was, as Ross Dunn points out, that Moulay Hassan's efforts failed largely because European powers refused to accept the Moroccan definition of the 'state' and because of the French army's refusal to accept Morocco's definitions of sovereignty (Dunn 1977; 169-170).

iii) From Figuig to the Dra

The Varnier proposals west of Figuig were not permanently accepted, however. After 1923, the French Protectorate in Morocco began to consider that a line somewhat to the south of the Varnier line would be more appropriate and, at the same time, that the centre of Ighli, at the junction of Guir and Zousfana rivers, should mark the southernmost extent of Moroccan sovereignty. The line then ran westwards, to the south of the Hammada du Guir and the Tafilalt, towards the Oued Dra which it joined at Taounite. A series of conferences were held during the 1920s between French officials in Morocco, Algeria and the AOF to resolve this issue and the others associated with Morocco's borders with the Western Sahara. However, they were unsuccessful in providing for any final delimitation and had to deal with a further series of disputes after 1921 reflecting differential customs duties between the Protectorate and Algeria. Further problems were caused by the fact that the putative dividing line crossed tribal migration routes (Trout 1969; 247-254).

In the end, the problem of the border region was rendered acute by continuing dissidence and, as a result, a new regime was created for it in 1929. This was the Confins Algero-Marocains, a zone of joint command between Algeria and Morocco, in which the Moroccan Resident General was to have overall military command, while the Algerian Governor General was to maintain administration of areas designated as Algerian. This system - which effectively avoided the thorny question of juridical sovereignty - lasted up to Moroccan independence in 1959. During the 1930s, however, there was a constant struggle between Algiers and Rabat over administrative responsibility for one portion of the Fonfins or another, once dissidence had been brought to an end (Trout 1969; 300).

By the Second World War, two conflicting views had emerged over the border, one held by the Protectorate and the other, with equal tenacity, by the Algerian authorities. The Protectorate claimed a line from the Jabal Grouz down the Oued Guir to just north of Ighli and then running as a straight line to contact the north eastern edge of the Western Sahara border along the $27^{o}40'$ N parallel just north west of Tindouf. The Algerian administration claimed a parallel line running north of the Protectorate line, which started from the Jabal Grouz and then west via the Jabal Mehiriz north of Bou Kais to run south along the escarpment of the Hammada du Guir at Djihani until it turned to run as a straight line to the Ktaoua bend of the Dra river from a point just north west of Abadla. It then followed the northern border of Spanish-held southern Morocco to the coast at Cap Noun - in a strange echo of the 1909 Niamey Conference line which set the southern Saharan borders of Algeria.

iv) The Trinquet line and the Limite operationnelle

By 1956, however, the administrative border for Morocco had become a line which was a compromise between the two extremes, although the starting point appears to have been at Ighli and the western limit that proposed by the Protectorate. At this point, an attempt was made to provide this administrative border with a historical pedigree reaching back to the 1930s, with a suggestion by the French Protectorate authorities that the true border was the so-called 'Trinquet' line which corresponded to the 1956.

Moroccan administrative border (Trout 1969; 415). The move, for which there appears to be no evidence before 1956, was clearly designed to provide a territorial limit for the newly independent state of Morocco, now shrunken compared with its pre-colonial extent but to be provided, nonetheless, with definitive territorial borders as befitted a successor state to the Protectorate which had been fashioned in a European mould.

The Trinquet line never acquired the status of an actual border, for events in the Algerian war of independence soon overtook these legalistic arrangements for Morocco's borders. Not only were there Algerian border posts north of the line, controlling the roads to the south of it, but from mid 1956 armed clashes along the line became increasingly frequent between the Moroccan Army of Liberation and French troops there. The confrontation continued until 1958, when French forces, in conjunction with Spanish troops in the Western Sahara, forced out the Moroccan liberation army unit, their auxiliaries and tribal supporters, mainly from the Reguibat Lagouacem. France set up a new boundary - the 'Limite operationnelle' - which ran from the Jabal Grouz, via the Jabal Zelmou, north of the Hamada du Guir and the Kem Kem plateau, to join up with the Dra river at Ktaoua. Morocco accepted, on May 23, 1958, that - for the time being - its forces would observe this delimitation, although it clearly did not accept that this was to be the frontier (Trout 1969; 425).

v) The 1963 border war and the 1972 treaty

Although Morocco agreed in 1961 that it would support Algerian territorial integrity against French attempts to separate off the Sahara - thus, incidently, abandoning its own claims to the Western Algerian Sahara which, in pre-Protectorate times, had been considered to be under Moroccan suzereinty and sovereignty, in return for an Algerian promise that the common border would be negotiable - events immediately after the end of the Algerian war of independence disrupted these understandings. In September, October and November 1963, Morocco and the newly independent state of Algeria fought a short border war along a front from Figuig to Tindouf. The war, which was brought to an end by OAU mediation, was eventually resolved by a ceasefire delimitation in February 1964, whereby Morocco was confined to the Dra oases and thus away from Tindouf, while Algeria withdrew from the Figuig and Ich.

The border issue remained unresolved for the next eight years, largely because Morocco could not abandon its claims - even to the Algerian Sahara in the aftermath of the war - while Algeria clearly enjoyed the support of the OAU, given the Cairo resolution on colonial borders, and would not compromise on its claim for the old Algerian administrative border. In 1972, in the

aftermath of two coup attempts against his regime and in the runup to Morocco's attempts to resolve the Western Sahara issue, Morocco and Algeria negotiated a final solution to their border problem (Brownlie 1979; 73-78). The treaty came in the wake of a 1969 treaty of solidarity and cooperation, signed at Ifrane in 1969 and the 1970 Tlemcen communique which set up a commission to delimit the border. The 1972 treaty implicitly accepted the inviolability of colonial frontiers and the pre-eminence of the original French Algerian administrative frontier running from the Dra to Figuig along what appears to be virtually the old 'Limite operationnelle'.

Although the June 15, 1972 treaty has not yet been ratified, it is virtually certain that, once other disputes between Algeria and Morocco have been resolved, particularly the Western Sahara issue, the treaty will eventually define the border. At the same time, it does not resolve the wider issues of the extent to which France failed to honour its obligations under the Algeceiras agreement to preserve the integrity of the Sharifian empire. The territories of Gouara, Touat and Tidikelt have been permanently lost to Morocco because of their original conquest by French forces under the command of French authorities based within the Algerian colonial administration. French administrative practice from 1901 onwards, after all, merely expressed administrative disputes between two different colonial regimes.

It is difficult, therefore, to accept that the OAU principle of the intangibility of colonial frontiers should be accepted here and Morocco - which did not accept the 1964 OAU resolution - has preferred to rest its case on paragraph 6 of the UN General Assembly resolution No 1514 (XV) of 1960 which, although it generally supports the usual UN practice for decolonisation, does allow that no decolonisation procedure should disrupt national unity or territorial integrity. Thus, although Morocco has been prepared to compromise on the principle for the sake of harmony within North Africa as far as its borders with Algeria are concerned, the principle itself has not been relinquished. After all, Gouara, Touat and Tidikelt were under Moroccan administration when France occupied them, the argument runs, and were thus under Moroccan sovereignty which has been outraged by the post-colonial settlement.

(b) Spanish imposed borders

The borders of Southern Morocco and the Western Sahara result entirely from colonial decisions by France and Spain. They are the consequence of France's desire to satisfy Spanish ambitions for some colonial presence in Morocco. As far as the western Sahara was concerned, Spain made it clear at the beginning of 1885 that it would exercise a protectorate over the Rio de Oro and this was confirmed and delimited by France, as far as the

AOF was concerned, in a convention in June 1900 (Brownlie 1969; 438-440). The northern section of the Western Sahara, the Saquiat al-Hamra, was subject to a further treaty between France and Spain in 1904 which laid down specific territorial limits - in the wake of the Entente Cordiale by which Britain renounced its own earlier interests in Morocco in favour of Morocco and, more specifically, in the Sahara region. The 1904 treaty acknowledged only that the Saquiat al-Hamra lay within a Spanish sphere of influence and it was only under a further treaty made in November 1912 that Spain acquired actual title (Brownlie 1969; 149-153).

Spain had also acquired title to protectorates inside Morocco itself by the 1904 and 1912 treaties. These included the famous northern zone of Morocco, the site of the Rif war from 1921 to 1926, the Tarfaya province in the south which formed the northern border of the Western Sahara and the enclave of Sidi Ifni just to the north of Tarfaya. Spain also claims title to the city enclaves of Ceuta and Melilla along Morocco's northern Mediterranean coastline, as a result of conquests in the fifteenth and sixteenth centuries. Most of these territories were abandoned in the wake of Morocco's accession to independence in 1956 - the Northern Zone in 1956, Tarfaya in 1958 and Sidi Ifni in 1969 (Knapp 1977; 334). Ceuta and Melilla still remain under Spanish control, but Morocco has made it clear that, eventually, they will have to be returned.

2.9 The Problem of Territorial Sovereignty

The issue as far as the old Spanish controlled areas of Morocco and the Western Sahara are concerned - as it has been with Mauritania, where Morocco was only prepared to recognise the state's independent existence in 1970 - is not one of border delimitation but one of sovereignty. Morocco does not necessarily dispute the colonial delimitations but does dispute the proposition that such colonially defined and created entities are entitled to an independent political existence. The formalistic basis for this rejection is the principle enunciated in paragraph 6 of the UNGA resolution 1514 (XV) of 1960. However, there is an underlying reality that goes far beyond this and confronts assumptions of territorial sovereignty that underlie modern precepts of international law and the principles of communal sovereignty that determined relationships between independent political entities in the pre-colonial period.

Modern concepts of territorial sovereignty derive from the situation in Europe in the mid seventeenth century as a result of the Peace of Westphalia, which brought the Thirty Years War to an end (Mazrui and Tidy 1984; 373-375). As a result of the treaty then created, rulers in Europe acquired sovereign power in the **territories** that they ruled. During the eighteenth and nineteenth centuries, this concept of a unique territory over which sovereign power could be exercised became associated with the additional concept of a single and unique **nation** which occupied it and, by the end of the eighteenth century, it had been accepted that sovereignty derived from the collective will of the nation situated on that unique sovereign territory (Joffé 1986; 8). Thus, the archetypal independent political unit in Europe became the **nation state,** with a homogeneous yet unique population defined by its cultural qualities, particularly language - the nation - living on a precisely defined and delimited territory that was uniquely its own - the state.

With the development of the industrial revolution, first in Britain and then in France and other European nations, the homogeneity of the nation was reinforced by an enforced population mobility which transferred rural populations to urban settings in order to service the nascent industrial base. The mobility itself, in a geographical sense, resulted from the radical change in occupation as agriculture ceased to be the mainstay of popular employment. At the same time, access to education and training began to change to allow for the changes in occupational activity and status which also incidently generated a sociological mobility. Thus the concept of nation - which developed as these other changes in social and economic organisation took place became associated with that of modernisation - itself the rallying cry of industrialisation, particularly in the nineteenth century. At the same time, European powers began a process of expansion which, in the case of France, brought it into contact and conflict, **inter alia,** with the diverse populations of North Africa where a different set of cultural assumptions held sway.

The colonial process, particularly in the case of France, was a process of cultural confrontation (Mazrui and Tidy 1984; 382). It was also at the same time, one of enforced dialogue, in that indigeneous opposition, once the initial period of primary resistance had given way to secondary resistance (Ranger 1968; 437-453), had to find a means of communicating its demands to the colonial authorities it faced. Even if the process was violent, eventually negotiation had to occur. However, for this to take place, there was a crucial need for the development of a common vocabulary in which the dialogue could be couched. Necessarily, however, the vocabulary would be dictated by colonialist preference, since the colonising power was the one which controlled the opportunities for discussion - and was also the one with the military power to ensure that dialogue could be denied or cut short if it did not follow acceptable patterns.

There was, however, usually another consequence - particularly if the nationalist opposition itself realised that dialogue was eventually inevitable, as was the case in North Africa. Since the opposition movement was inherently modernist in its outlook, it tended to incorporate much of the philosophical assumptions of the colonising power into its own ideology - particularly those which dealt with the formation of a modern nation state. It often therefore considered a nationalist ideology to be an essential part of the ideology that it would use to confront the colonising power in any dialogue. This meant that, inevitably, indigeneous nationalist movements mobilised around ideologies that owed much to their colonial precursors and that the states they later created owed much to the territorial constructs of colonial times.

This tendency was particularly significant in North Africa, largely because two of the modern states there had no pre-colonial precursors while the other two now have forms very different from those of pre-colonial times. There was also the tradition of an ideological precursor which made use of similar methods of mobilisation, albeit on a different scale and for different purposes - Islam. This made acceptance of nationalist ideologies which owe much to European precursors easier to accept, particularly amongst the population at large and particularly if the ideology itself was expressed in terms that related to more traditional political symbols (Joffé 1985a; 307). Nonetheless, the colonial legacy has generally obliged successor states to attempt to justify colonial administrative creations in terms of a nation that is assumed to occupy the state that has thus been created - even if the concept of a 'nation' had been alien or irrelevant to the precolonial situation. Over time, of course, nations have emerged in response to the immense efforts of independent governments to weld a political consensus out of often disparate colonial and precolonial sociological legacies. Nonetheless, the precolonial, political and sociological reality still persists and the 'nations' thus formed are often uncertain creations. Equally, assertion of the existence of such nations as a justification of territorial sovereignty is often open to dispute - as has been the case in the Western Sahara.

Libya, for instance, has acquired a political and territorial unity than not even the Ottomans or the Sanusi could create. Algeria now covers a territorial extent that no pre-colonial ruler could ever have imagined. Morocco, on the other hand, is truncated, having lost the Saharan provinces which it clearly controlled in the ninteenth century to Algeria, yet now defiantly justifying a 'recovery' of the Western Sahara that it claims corresponds to its communal links in the pre-colonial era. Only Tunisia, with its Saharan extension, has avoided major problems over its territorial claims and control. In each case, the state concerned has also made great efforts to describe its population in terms of a fully integrated nation to justify its territorial claims and has reacted very strongly to any example of particularistic sentiment (Joffé 1985b; 39-49).

The justification used by each state for its right to the territory it now controls or claims is also in keeping with its adoption of European values and definitions of the concept of 'state'. Algeria, which sprang new born from the colonial era, justifies its territorial extent and its borders by the twin principles of the inviolability of colonial frontiers as laid down by the OAU and of **uti possidetis** in that it is the successor to a colonial entity created by France. Libya claims the same, adding only that its justification derives not only from the Italian experience but also from the prior Ottoman occupation. In addition, it would claim that the French administrative record, on which Algeria bases its case for the Ghat region, is defective and that the true extent of **uti possidetis** is different from that argued by Algeria. In reality, however, behind the Libyan claim lurks an earlier vision - that of communal sovereignty and the **umma.** The Sanusi, after all, extended control south to Northern Chad and traditional economic links brought the Fezzan and the Tebu together.

Morocco, too, would argue a very similar case even more forcefully over the Western Sahara. Here, after all, the descendents of the original Moroccan conquerers of Timbucktoo, the **arma**, continued to elect their own pashas until the 1790s while recognising an overall Moroccan suzereinty throughout (Trout 1969; 144). Furthermore, the Ma al-Ainin family, in the Saquiat al-Hamra, was formally linked to the Moroccan sultanate during the reign of Moulay Abd al-Aziz at the start of the twentieth century and also accepted Moroccan suzereinty (Trout 1969; 156-160). International treaties, too, also lend credence to this view (Trout 1969; 204), although it must also be noted that the International Court of Justice was very cautious in refusing to consider that these links actually amounted to full **territorial** sovereignty when it investigated Morocco's claims in 1974 (ICJ 1974).

Sovereignty of Morocco

2.10 The Consequences

Indeed, the Moroccan case underlines the fact that these concepts of sovereignty that now inform the situation in independent North Africa are articulated with all the force and precision of territorial delimitation that is typical of the European experience, even though the underlying principle is one of communal sovereignty, in which territory was not of primordial importance. The explanation is simply that the successor states to the colonial legacy are obliged by the logic of the ideologies they have incorporated through the experience of national liberation to claim a nationalist heritage within their sovereign territories. Thus, ironically, border disputes in North Africa are the ultimate colonial legacy in that they have forced the independent states of North Africa to adopt that typical form of European political organisation, the nation state. In doing so, they have also accepted the primacy of precise territorial delimitation in defining their uniqueness and sovereignty.

2.11 References

Abun Nasr J. (1975) **A history of the Maghrib,** CUP

AGGA - **Archives of the Government General of Algeria;** Aix-en-Provence

Bougaita B. (1979) **Les frontières méridionales de l'Algérie,** Thesis, Paris

Brignon J. **et al** (1967) **Histoire du Maroc,** Hatier

Brown C.L. (1974) **The Tunisia of Ahmad Bey: 1837-1855,** Princeton UP

Brownlie I. (1979) **African boundaries, a legal and diplomatic encyclopaedia,** Hurst/RIIA

Decalo S. (1977) **Historical dictionary of Chad,** Scarecrow Press

Delupis I. (1974) **International law and the independent state,** Crane Russak

Dunn R. (1977) **Resistance in the desert,** Croom Helm

Goodchild R.G. (1976) **Libyan Studies,** Elek

Hernia A. (1980) **Le Grid: ses rapports avec** le **Beylik** de **Tunis (1676-1840),** Publications de l'Universite de Tunis

Hodges A. (1984) **Western Sahara: roots of a desert war,** Croom Helm

ICJ (1974), International Court of Justice (1974) **Western Sahara,**

Joffé E.G.H. (1981) 'Libya and Chad', **ROAPE,** 21, May-September 1981

ibid (1982) 'Social and political structures in the Jefara plain in the late nineteenth century', in Joffe E.G.H. & McLachlan K.S. (eds), (1982), **Social and economic development of Libya,** Menas Press

ibid (1984) 'The French occupation of the Western Jefara plain and the village of Dahibat - 1890-1981', **Libyan Studies, 15,** 1984

ibid (1985a) 'Nacionalismo y el mito democratico: la experiencia norteafricana', **Cuadernos de Alzate, l,** Invierno 1984-1985

ibid (1985b) 'The Moroccan nationalist movement: Istiqlal, the sultan and the country', **JAH, 26** (1985)

ibid (1986) 'The International Court of Justice and the Western Sahara dispute', paper for the Western Sahara Conference, Oxford, March 1986

Knapp W. (1977) **North West Africa, a political and economic survey,** OUP

Lahbabi M. (1968) **Le gouvernement marocain à l'aube de XXe siècle,** Editions Atlantides

Lanne B. (1982) **Tchad-Libye: la querelle des frontières,** Karthala

Mardin S. (1962) **The genesis of young Ottoman thought: a study in the modernisation of Turkish political ideas,** Princeton

Martel A. (1965) **Les frontières Saharo-Tripolitaines de** la **Tunisie,** PUF

Mazrui A.A. and Tidy M. (1984) **Nationalism and new states in Africa,** Heinemann

Morsy M. (1984) **North Africa 1800-1900: a survey from the Nile valley to the Atlantic,** Longmans

Muller M. (1982) 'Frontiers, an imported concept', in Allan J.A. (ed) **Libya since independence,** Croom Helm

Murty T.S. (1978) **Frontiers, a changing concept,** Palit & Palit

Nelson H.D. (1979) **Morocco, a country study,** American University

Pennell C.R. (1986) **A country with a government and a flag; the Rif War in Morocco,** Menas Press

Pitcher D.E. (1971) **An historical geography of the Ottoman empire,** Brill

Prescott J.R.V. (1978) **Boundaries and frontiers,** Croom Helm

Ranger T.O. (1968) 'Connections between 'primary resistance' movements and modern mass nationalism in East and Central Africa, part 1, **JAH,** IX

Schacht J. (1964) **An introduction to Islamic law,** OUP

Stuart G.H. (1931) **The international city of Tangier,** OUP

Thomas C. (1985) **New states, sovereignty and intervention,** Gower

Trout F.E. (1969) **Morocco's Saharan frontiers,** Droz

Wright J. (1969) **Libya,** Benn

Yahya D. (1981), **Morocco in the sixteenth century,** Longmans

Zartman W.I. (1984) 'Maghrebi politics and Mediterranean implications', in Luciani G. (1984) **The Mediterranean region,** Croom Helm

FIG 1. Borders of Algeria

FIG 2. Borders of Libya and Chad

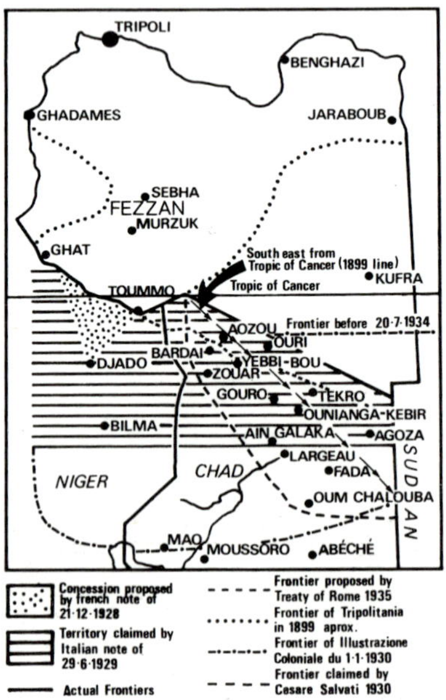

FIG 3. Borders in the Western Sahara Area

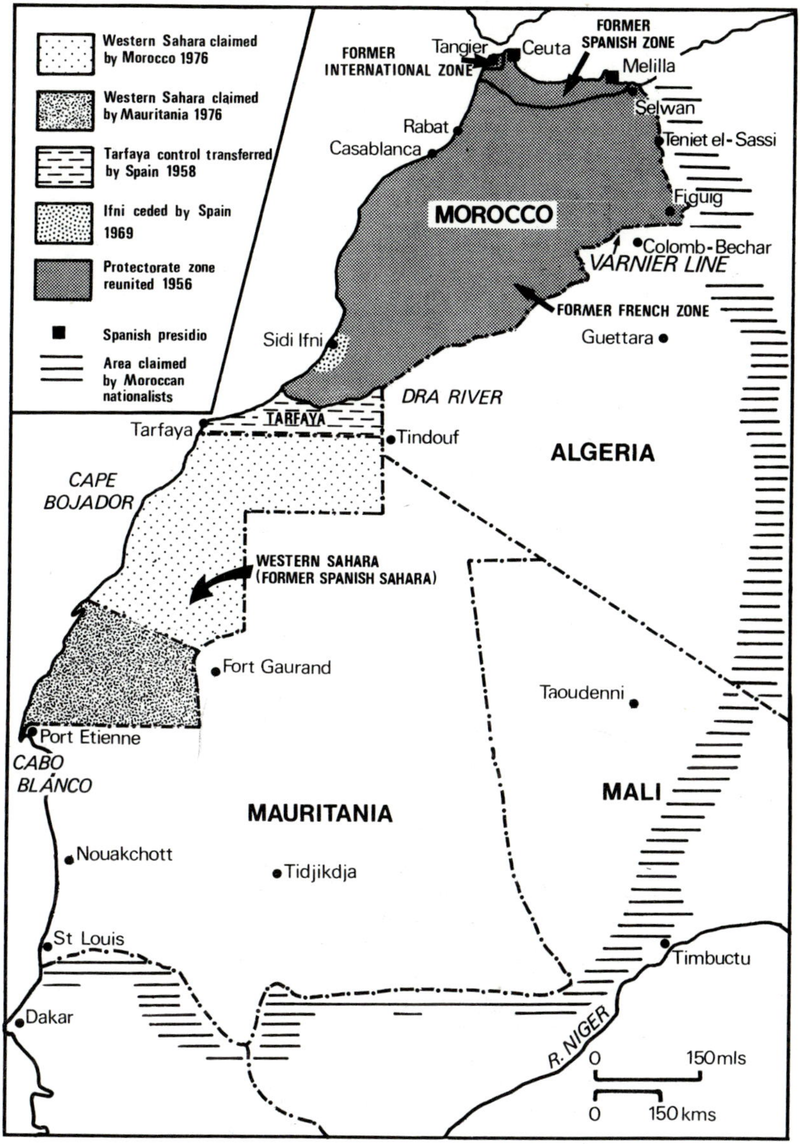

3. THE DEVELOPMENT OF THE EGYPTIAN-ISRAELI BOUNDARIES 1906-1986

Nurit Kliot

3.1 Introduction

This paper examines the development of the Egyptian-Israeli boundary between 1906, when it was first allocated, and 1986. The first section investigates the development of the boundary between 1906 and 1914 and the specific stages of allocation, delimitation, and demarcation of the boundary. The second section discusses the development of the boundary from the year 1914 to 1947, while the third section examines the evolution of the boundary between 1948 and 1986, after Egypt and Israel had fought each other in four major wars which ended with the 1979 Peace Treaty.

3.2 The Development of the Boundary: 1906-1914

On October 1, 1906 the separating administrative line between the Ottoman province of the Hejaz, the governorate of Jerusalem and the Sinai Peninsula was agreed upon, following nine months of military action and diplomatic activity which almost brought the British and Ottoman empires to the verge of war. This boundary became, with minor changes, the present boundary separating Egypt and Israel.

The evolution of that particular boundary has been well documented in both Turkish and British sources. Moreover, two of these sources are the accounts of the two men on the spot: Rushdi Pasha, the Turkish commandant of Aqaba, and W. Jennings-Bramly, the British frontier administration officer in Sinai in 1910/11. Rushdi Pasha published a detailed account of the events in "Aqaba Messelhessi" (The Aqaba Problem), which was published in Turkish. W. Jennings-Bramly never published his memoirs but in 1959 left all his papers to the Royal Geographical Society in London. A third source consists of the official correspondence of the British Foreign Office and the Cabinet papers of the years 1905-1907, regarding the frontier dispute. This availability of varied sources has enabled various researchers to analyse, in detail, the historical development of this boundary (see, for instance, Brawer 1970; Heyd 1963; Warburg 1979).

3.3 The Allocation of the Egyptian-Israeli Boundary: 1906

Allocation refers to the general political decision on the division of territory. The allocation process of the Egypt-Palestine border was imposed on the Turks by the British in 1906. The first map to indicate the formation of modern Egypt appeared in 1841 in the Sultan's firman which showed the boundaries of the area under Muhammed Ali's rule. This map showed that the boundary between Egypt, Hejaz and Palestine ran from Suez toward Rafah and then eastward (Biger 1978, 325). It also showed that the Sinai peninsula remained under Hejaz (Turkish) administration and Palestine ended in the Rafah-Dead Sea Line **(see Fig. 3.1).**

3.4 British-Turkish Confrontation of 1906

British administrators who had ruled Egypt from 1882 were anxious in 1906 to demarcate the boundary between Egypt and Palestine. The incident which led to the confrontation between the British and the Turks took place on January 10 1906. W. Jennings-Bramly was ordered by Sir Reginald Wingate, then Governor of Sudan and Sirdar of the Egyptian army, to proceed from his headquarters in Nakhl to Umm Rashrash (now Eilat) to build up a small post for the Egyptian border police. (The Jennings-Bramly papers, Royal Geographical Society, London).

Bramly, accompanied by four or five Egyptian policemen, asked the Turkish officer at Aqaba to confirm that he had been ordered by the Sirdar to establish border posts in the western parts of the Gulf of Aqaba, parts of the southern Negev, Kuntila and Kusseima. Bramly claimed that all these parts belonged to Egypt and that he had therefore to organize a border guard for it (Heyd 1963, p.196; Warburg 1979, 679). Rushdi Pasha, the local Turkish officer, travelled to Syria in order to establish if Umm Rashrash was within the Turkish or the Egyptian administrative boundaries. When he came back, he ordered Bramly and his men to leave Umm Rashrash immediately (Heyd 1963). Bramly returned to his headquarters in Nakhl and the Turks founded a Turkish border post in Umm-Rashrash.

The second stage of the incident began on January 22, when a small navy vessel, the Nur al-Bahr, anchored near Jezirat-Farun. Bramly himself was ordered to return to the Gulf of

Aqaba. Bramly told Rushdi Pasha that he had orders to post 50 Egyptian soldiers in Taba and Umm-Rashrash. Rushdi Pasha, who had succeeded in posting Turkish troops in Taba the night before, told Bramly that he had been ordered by the Ottoman Grand Vizier to oppose the Egyptian occupation of Taba by force if necessary (Heyd 1963, 197). In February 1906 the Turkish government sent a large reinforcement of troops to Aqaba and the number of Turkish troops reached 3,000. The events of subsequent months brought about little change. A British gun boat, the Diana, was rushed to the Gulf of Aqaba, while Turkey further reinforced its troops in Aqaba. Britain retaliated in May by rushing additional forces to Egypt and by ordering its Mediterranean fleet to concentrate in Piraeus. In April a British attempt to land troops in Rafah, the northern tip of the Turco-Egyptian border, was opposed by the Turks and had to be abandoned (Warburg 1979, 680).

Britain now resorted to diplomacy backed by force. Egypt demanded a demarcated line between the Egyptian and Turkish territories but the Sultan rejected this demand, claiming that Egypt was an integral part of the Ottoman Empire and that there was therefore no question of a formal boundary separating two sovereign states. Under British pressure, the Turks first suggested that Egypt would administer two areas in the Sinai: the first, east of the Suez Canal, up to the line leading from Rafah to Suez, corresponded to the area granted to Muhammad Ali in the firman of 1841; the second part was the southern section of the peninsula up to a line in the north, leading from a few kilometres west of Aqaba to Suez. This suggestion was opposed by Britain **(see Fig. 3.1)**. A second compromise was proposed by Turkey, dividing the peninsula into two halves, the separating line leading from El-Arish in the north to Ras-Muhammad in the south. This proposal, which left the entire western half of Sinai, including the Suez Canal and its approaches, under Egyptian administration, was also rejected by Britain (Cromer 1908, Vol.11, 269).

On May 3 1906 Britain, supported by France and Russia, handed an ultimatum to Sultan Abdulhamid 11 warning him that unless he agreed within ten days to the proposed boundary leading from Rafah to the Gulf of Aqaba, and evacuated Taba, His Majesty's Government would be forced to take stiff measures (Gooch and Temperley 1928, Vol, V 190-1). It is known today that at the same time Britain was making preparations to capture a few important Turkish islands in the Aegean Sea. On May 14 the Sultan, realizing that without German, Russian or French support he could not afford to resist, accepted the British demand and ordered the withdrawal from Taba and the delimitation of the administrative separating line between the Ottoman province of the Hejaz, the governorate of Jerusalem and the Sinai Peninsula

(Warburg 1979; 681).

3.5 Background to the Turkish and British Positions in the Dispute

The most important question concerning the British-Turkish confrontation is whether the British or the Turks, or both, decided at the end of 1905 to expand their rule in the Gulf of Aqaba and eastern Sinai. There is some evidence for a British initiative in Cromer's report on Egypt (Cromer 1906; 268-9). It seems that by the end of 1905 the British had decided to show a stronger military presence in Sinai. But it is also possible that the Turks were interested in imposing their sovereignty over the Sinai peninsula, after their failure to do so in 1892. When Sultan Abdulhamid II decided in 1892 to re-establish his authority in the region belonging to Turkey, as drawn on the 1841 map, he did so on the advice of the German Ambassador to Turkey. Heyd (1963) suggested also that the 1906 British-Turkish dispute was part of the Anglo-German competition for hegemony in the Middle East and elsewhere.

The particular considerations of British and Turkish positions in relation to the exact location of the boundary should be specified. The Turks wished to leave the central triangle of Rafah - Aqaba - Suez in Turkish territory, enabling the Turks to extend the Hejaz railway from Ma'an to Aqaba and from there to Suez. The fall of 1904 saw the opening of the Hejaz railway line from Syria to Ma'an and the Turks were able to transfer troops to Ma'an and from there, through Aqaba, to Yemen. There were rumours that Turkey was planning to connect Ma'an to Aqaba by railway,a link which would free it from dependence on the British-controlled Suez Canal (Graves, quoted in Heyd 1963, 203). The British were concerned about this Turkish plan. They feared that a Turkish military base in Aqaba would endanger the sea-route from Suez to India and seriously increase Egypt's vulnerability (British documents, Vol. V, 39). By drawing the boundary between Egypt and Palestine on the Rafah-Aqaba line, the British made sure that the Turks would be located as far as possible from the Suez Canal. In addition, the Sinai desert presented a natural barrier and good defence for the Canal. Control over the Sinai Peninsula also meant supervision of any military force in Eastern Sinai or the Western Negev (Brawer 1979, 369-370). This British position appeared in Lord Curzon's writings quoted in Brawer (1970; 127). According to Brawer, the British followed the opinions and experience of Napoleon when they demanded that the boundary begin at Rafah and not in El-Arish, the traditional gateway to Sinai. By putting the boundary at Rafah, the British further distanced the Turks from the Suez region (Brawer 1970; 128).

3.6 The Delimitation of the Egypt-Palestine Boundary

Delimitation is defined as the final selection of a specific boundary within what is frequently a broad allocation zone. After the allocation of the Egyptian-Palestinian border had been determined as the "Rafah-Aqaba line", the British approached the second stage of actually delimiting the boundary. A special British Commission studied the geographical features of the Rafah-Aqaba zone. The geographical study of the border zone included a detailed survey of all water resources, their quantity and quality. A second feature surveyed by the British were the main roads of eastern Sinai, the third the Bedouin tribes living in the border zone. The British Commission made a detailed survey of the location of the different tribes, the lands they cultivated, their grazing areas, their movements and their affiliation to different centres. Finally, the British surveyed the geomorphology and topography of the border zone. On the bases of all four features, Jennings-Bramly suggested a "natural boundary".

This began on the Mediterranean coast north-west of Rafah, continued straight south for 16 km then turned west up to Wadi El-Arish. From Wadi El-Arish near Bir-Lachfan the boundary turned south-east along the Wadi up to its connection with Wadi Moilech and Wadi El-Ein **(see Fig. 3.1)**. This boundary line was based on natural geographical features such as wadis and mountains and also preserved the grazing areas of the various tribes. But this boundary line was meandering and complex and the British Command rejected Bramly's boundary. They were not concerned about dividing tribal territories or about the necessity of a boundary based on natural features. Another important aspect of Bramly's boundary was that it included large parts of the Arava (Rift Valley) within the Egyptian side of the border and the British knew that the Turks would never agree to such a major change in the border. The most important consideration for the British authorities was that the boundary should be short, perhaps a straight line with both strategic military and political advantages (Brawer 1979; 372). Consequently, the final boundary ignored many of Jennings-Bramly's proposals but still provided the British with all the advantages they had been looking for: almost all the available water resources were included in the Egyptian side of the border; all the major roads and important junctions in eastern Sinai remained on that side (Brawer 1970; 131-132). The boundary did cut across tribal areas and many of the tribes had to move to one side of the border or the other.

3.7 The Demarcation of the Egyptian-Palestinian Boundary

Demarcation is defined as the actual marking of the boundary location by posts and other markers. The Ottoman and British Empires finally agreed upon a straight boundary line from Rafah to Aqaba. In actual fact, the boundary deviated from the agreed line because of the region's harsh topography and hot climate which were to make the final demarcation a very difficult exercise. Fourteen permanent telegraph poles, which were replaced later by stone posts, were laid in cement foundations. The boundary started at Ras-Taba (10 m south west of Aqaba) and not at Umm-Rashrash. This shift in the boundary location is considered a Turkish victory resulting from some serious bargaining. In other parts of the boundary the deviations from the agreed straight line were a direct outcome of the steep mountains and deep canyons inaccessible to the British surveyors. For example, in the Harif mountain the boundary was mistakenly demarcated 500 metres east of its proper site. Basically, the boundary remained a straight geometric line even when it was in total contradiction to the topographical elements **(see Fig. 3.1).**

3.8 Some Concluding Remarks on the Development of the Border in 1906

The allocation, delimitation and demarcation of the Egyptian-Palestinian border show two important features. First, Turkey probably lost its direct control over Sinai because it entirely failed to impose its sovereignty on Sinai. It did not establish an administration and had no military forces there. The British, seeing this vacuum, took advantage of the opportunity to expand their control over the area. Jennings-Bramly stated that between 1841 and 1905 there had been no change in the status of Sinai and that Turkey, not having acted to establish its sovereignty, had to surrender to British threats (Jennings-Bramly 1904, Document 1, Brawer 1979; 370).

3.9 The Development of the Egyptian-Palestinian Borders: 1914-1947

In October 1906, the final demarcation of the boundary between Palestine and Egypt was completed, but until the First World War the new boundary had almost no impact on either country or on their border inhabitants. Very few military posts, with their strictly symbolic presence, were located near the new boundary - mainly on its western side, and the Bedouin in the border area continued their nomadic crossings of the border as if it did not exist (Brawer, 1979; 373). Following the declaration of a British protectorate over Egypt in December 1914, Britain could no longer disguise the fact that it was her own interests, the Suez Canal, the protection of her route to India and her general strategic policy, which were at stake (Warburg 1979; 683). With the dismembering of Turkey the problem of Sinai was no longer a Turco-Egyptian problem. The old Separating Administrative Line became a border between British-controlled Egypt on the one hand and British-controlled Palestine and Trans-Jordan on the other. In the years 1918-1922 the British government held major discussions regarding the exact demarcation of this border. The final decision was to leave the 1906 administrative separating line intact while the question of sovereignty (i.e. Egyptian sovereignty over Sinai) remained in abeyance. Some British Foreign Office papers pointed to the fact that at least some officials regarded the old separating line of Rafah-Aqaba as in effect an administrative division between the two Ottoman provinces (Fischer-Williams, 1926 quoted in Warburg 1979; 684). Following the Second World War the question of the boundaries of Egypt in the Sinai peninsula was raised once again by Jennings-Bramly. According to Bramly, the 1906 agreement did not settle the legal question of the Egyptian boundary in Sinai. He suggested that Great Britain should claim suzerainty in that area by right of conquest. He held that Sinai was either Turkish or "no man's land" but in no way Egyptian. The Foreign Office rejected Bramly's proposal, stating that since the First World War the whole of the Sinai peninsula had been accepted by Britain and other states as Egyptian territory. But the Foreign Office officials admitted that neither the 1923 Treaty of Lausanne nor the declaration of a British protectorate over Egypt in 1914 had in any way bonded the Sinai peninsula legally to Egypt. In 1947 the British Foreign Office again took up the question of the legal ownership of Sinai, and again produced two different views: one which claimed that there was no legal owner to Sinai and a diametrically opposed view (Howson 1947, quoted in Warburg 1979, 687; Mervyn Jones, 1947 quoted in Warburg 1979, 687).

3.10 The Development of the Israeli-Egyptian Borders: 1948-1986

Between 1920 and 1947 Palestine was a British mandate. Following continuous conflict between Jews and Arabs in Palestine, the UN proposed a plan partitioning the country into a Jewish and an Arab state (Drysdale and Blake, 1985; 279). It is important to note that during discussions on the future boundaries of the Jewish state the Americans objected to including the Negev or most of it in the areas allotted to this state. The real issue was of course the old one of British control of the Suez Canal. The British did not want anything separating their bases in the

Suez Canal zone from those in Trans-Jordan (Cohen, 1970; 388). Furthermore, they considered the Negev an alternative base, were they ever forced to evacuate the canal zone. The Soviet Union opposed the reduction of the proposed Jewish state and the exclusion of the Negev from its territory, as suggested by the US representative.

The Arab countries, fiercely objecting to the partition plan, intensified their attacks on Jewish settlements and on May 15, 1948, one day after Israel declared its independence, the armies of its Arab neighbours attacked Israel. Jewish settlements in the Negev were attacked by the Egyptian army in May 1948. Toward the end of the war, Israeli forces captured the whole Negev except the Gaza Strip. The Israelis' pursuit of the Egyptian forces brought them into Sinai and they captured a vast section if it between El Arish and Nitzana (Map 2) (Lorch, 1976; 75; Cohen 1970; 401-2). Resulting from the Armistice Agreements signed by Israel and Egypt on the 24 February 1949, Israel retreated from the area it occupied in Sinai, Egypt acknowledged Israels possession of the entire Negev (except for the Gaza Strip) and the Auja-al Hafir (Nitzana) area was demilitarized and made a headquarters of the mixed armistice commission (see map 2). Whereas the partition plan allotted to Israel an area of 15,850 sq km, over 9,500 sq km of this in the Negev, at the end of the 1948-9 Israeli War of Independence Israel held 20,770 sq km; the Negev comprises two-thirds of the area of the state.

The boundary between Israel and Egypt did not become a peaceful boundary when the war terminated and armistice agreements were signed. From the time of the armistice up to the Sinai Campaign (October 1956) there had been 11,650 incidents on the Israeli-Egyptian border (Cohen 1970; 467). In the period 1951-6 over 400 Israelis were killed and 900 injured as a result of Arab infiltrations and attacks (Lorch, 1976; 83). Most of the early incidents in the early 1950's sprang from innocent motives: Arab refugees trying to return to their villages, Bedouin crossing the border with their animals, accidental border crossings by both Arabs and Israelis because the demarcation lines were often marked inadequately or not at all (Khouri, 1968; 183-4). But sabotage, thefts, terror increased in number and frequency and Israel retaliated, especially in 1955 and 1956 when Fedayeen attacks on Israel intensified. Between 1 January 1955 and September 1956, according to UNTSO sources, 496 Arabs were killed and 419 injured in these attacks (Khouri, 1968; 188). Another source of tension between the two states was the closing of the Straits of Tiran and the Gulf of Aqaba (and the Suez Canal) to Israeli shipping. In 1956 Israel cooperated with France and Britain in the Sinai Campaign. This was the last time when British control of the Suez Canal guided British and French geopolitical policy. Britain and France felt threatened by Nasser's nationalization of the Canal; Israel wanted a secure border with Egypt and freedom of shipping to Israeli ships in the Gulf of Aqaba and the Suez Canal. Israeli gains included freedom of shipping in the Straits of Tiran and Gulf of Aqaba and cessation of the Fedayeen attacks from the Gaza Strip through the stationing of a UN force along the border in Sinai as well as in the Sharm el-Sheikh area. But Israeli shipping was still denied the use of the Suez Canal (Khouri, 1968; 218).

Israel embarked on a policy of phased withdrawal from Sinai, coupled with efforts to wrest some political concessions from Egypt and the super-powers, just as it did between 1976-1982. Israel occupied Sinai from October 29th 1956 until 22nd January 1957, when it returned most of Sinai, except the Gaza Strip and Sharm el-Sheikh which were evacuated on March 6-10 1957, **(see Fig. 3.2).** For 10 years relative calm prevailed along the border, and as a result of freedom of navigation for Israeli shipping Eilat became an important port for Israeli trade with East and Southern Africa and Asia and also as a crucial port for Israel's oil imports (Downing and Herman, 1978; 75). From 1965 on, attacks against Israel resumed and Israel retaliated. In 1967, Egypt demanded that the U.N. evacuate its peace-keeping forces from Sinai and announced the closing of the Straits of Tiran and the Gulf of Aqaba to Israeli shipping. This resulted in the Six Day War from 5 June to 10th June, 1967, when Israel occupied Sinai, the Jordanian West Bank and the Golan Heights **(see Fig. 3.2).** Between 1968 and 1970 Israel and Egypt fought on the Suez front for more than 1,000 days but there were no territorial changes and Israel continued its occupation of Sinai though both sides suffered heavy casualties.

In October 1973, the fourth war between Israel and Egypt began with a surprise attack on Israel on both the Syrian and the Egyptian fronts. When the ceasefire agreement was signed on the Egyptian front, Israel was occupying 1,600 sq km west of the Suez Canal - inside Sinai **(see Fig. 3.2).** The Yom Kippur War and its final stalemate eventually brought the two countries to sign a peace treaty. But even before this Israel withdrew its forces in a series of military agreements between the two countries. In 1974, Israel and Egypt signed an accord to separate their military forces. Another disengagement agreement was signed in September 1975. Israel withdrew from part of the territory which she had occupied and UN forces were placed in buffer zones between the Israeli and Egyptian forces. In the second disengagement agreement (called the Interim Agreement), Israel withdrew from the strategically important Gidi and Mitla passes and from the Abu Rudeis oilfields **(see Fig. 3.3).** Egypt for the first time agreed "not to resort to the threat or use of force or military blockade against Israel" and non-military cargoes destined for or coming from Israel were permitted through the

Suez Canal (Akehurst 1981, p.1038, Cantori 1984; 173). Between 1975 and 1979 the peace negotiations between the two countries continued, culminating in the peace treaty.

Two months after its formal inauguration (April 25, 1979) Israel restored to Egypt the town of El-Arish, the administrative capital of Sinai. By 25 November 1979 the Alma oilfields had been returned, and by 25 January 1980 two-thirds of Sinai (Map 3). On the latter date, the land borders with Israel were opened. Finally on 25 April 1982 the remainder of Sinai was turned over to the Egyptians.

The Egyptian-Israeli peace treaty made special provisions for the demilitarization of both Sinai and a narrow strip of Israeli territory. Sinai was divided into three zones **(see Fig. 3.4).** In zone C, which ran parallel to the international border and the Egyptian shore of the Gulf of Aqaba, no Egyptian forces were to be stationed. The central zone B, with an average width of 100 km, was to contain no more than 4,000 lightly armed Egyptian troops. In the westernmost zone A, Egypt was limited to 22,000 armed men and a specified numbers of weapons and armoured vehicles **(see Fig. 3.4).** As a reciprocal measure a narrow zone D, extending approximately 5 km, was established on the Israeli side of the international boundary in which the Israel Defence Force was limited to 4,000 persons, with further limitations on the number and nature of their weapons and vehicles (James 1981; 255: Sobel 1980; 257). A multi-national force and observers comprised of 2,600 military personnel from ten countries supervises the implementation of the demilitarization arrangement in Sinai and free navigation in the Straits of Tiran.

Finally, according to the Peace Treaty, the boundary between Egypt and Israel was delimited and demarcated everywhere except on its southernmost point at Taba where an area of 900 square metres is disputed. Egypt and Israel are disputed on the exact location of the boundary near Taba and on the way to solve this dispute. In September 1986 the two countries signed the compromise of Taba arbitration and agreed on the three people who will serve as arbitrators, thus resolving the dispute according to the provisions of the Peace Treaty.

3.11 Summary and Conclusions

The allocation, delimitation and demarcation of the Egyptian-Israeli boundary was determined by the power struggle between two empires: the British and the Ottoman. In this struggle the British won and the boundary was allocated and delimited to satisfy their geostrategic needs. During the eighty years of its existence, the Egyptian-Palestinian (later Israeli) boundary gained stability and permanence although there were legal doubts about the Egyptian right of sovereignty over Sinai. This stability was reinforced by the behaviour of Israel, which fought with Egypt in 1948, 1956, 1967 and 1973 but which returned to the former boundary after each of the above wars (except for the years 1967-1979 in which Israel occupied Sinai). The 1979 Peace Treaty between Israel and Egypt has finalized the Rafah-Eilat boundary though there are still minor disputed areas along the border such as Taba which have not yet been demarcated and on which the two countries are still in dispute.

3.12 Acknowledgements

This paper was written while the author was on sabbatical leave at the Department of Geography, the London School of Economics. The author would like to express her acknowledgement to the Department of Geography for its assistance.

References

Akehurst, M. (1981), 'The Peace Treaty between Egypt and Israel'., **International Relations,** V11, 1, May, pp. 1035-1052.

Biger, G. (1978), 'The First Map of Modern Egypt: Mohammed Ali's Firman and the Map of 1841'. **Middle Eastern Studies,** 14, pp.324-325.

Brawer, M. (1970), 'Geographical Elements in the Formation of the Egyptian-Palestine Boundary'. in **Studies in the Geography of Israel,** VII, pp. 125-37. (in Hebrew).

Brawer, M. (1979), **Boundaries and Frontier in the Negev,** in A. Shmueli and Y. Gradus (eds), **Land of the Negev,** University of Tel Aviv, Tel Aviv, pp.368-379. (in Hebrew).

British and Foreign State Papers, 99, pp.482-84.

Cantori, L.J. (1984), **Egyptian Policy,** in R. Freedman (ed), **The Middle East Since Camp David,** Boulder, Colorado: Westview, pp.171-191.

Cohen, A. (1970), **Israel and the Arab World,** London: W.H. Allen.

Cromer, Lord (1908) **Modern Egypt,** Vol.II, London, pp.268-9.

Curzon, Lord Kedleston (1907), **Frontiers,** Oxford University Press, Oxford.

Downing, D. and Herman G., (1978), **War without End - Peace without Hope,** London, New English Library.

Drysdale, A. and Blake G.H. (1985), **The Middle East and North Africa - A Political Geography.** New York, Oxford University Press.

Gooch, G.P. and Temperley, H. (eds) (1928), **British Documents on the Origins of the War 1898-1914,** V, pp.190-1.

Heyd, U. (1963), **The Gulf of Eilat Crisis of 1906** in Eilat - The Annual Meeting of the Society for the Study of Eretz-Israel, Jerusalem, pp.194-206. (in Hebrew).

James, A. (1981), 'Symbol in Sinai: the Multinational Force and Observers', **Millennium Journal of International Studies,** 14, 3, p.255-276.

Jennings-Bramly, W.E. (1906), Collection of documents **Relating to the Dispute over the Turco-Egyptian Frontier,** Royal Geographical Society, London, (Documents 1 and 2).

Khouri, F.J. (1968), The **Arab-Israeli Dilemma,** Syracuse, New York, Syracuse University Press.

Lapidot, R. (1985), 'The Problem of Taba and Solving Disputes Peacefully', **Seqirah Hodshit,** 6, pp. 13-17. (in Hebrew).

Lorch, N. (1976), **One Long War,** Jerusalem, Keter Books.

Marshall-Cornwall, J. (1959), 'An Enigmatic Frontier', **The Geographical Journal,** CXXV, pp. 459-62.

Sobel, L.A. (ed) (1980), **Peace Making in the Middle East,** New York, Fact on File Inc.

Warburg, R. (1979), 'The Sinai Peninsula Borders, 1906-47', **Journal of Contemporary History,** 14, pp.677-692.

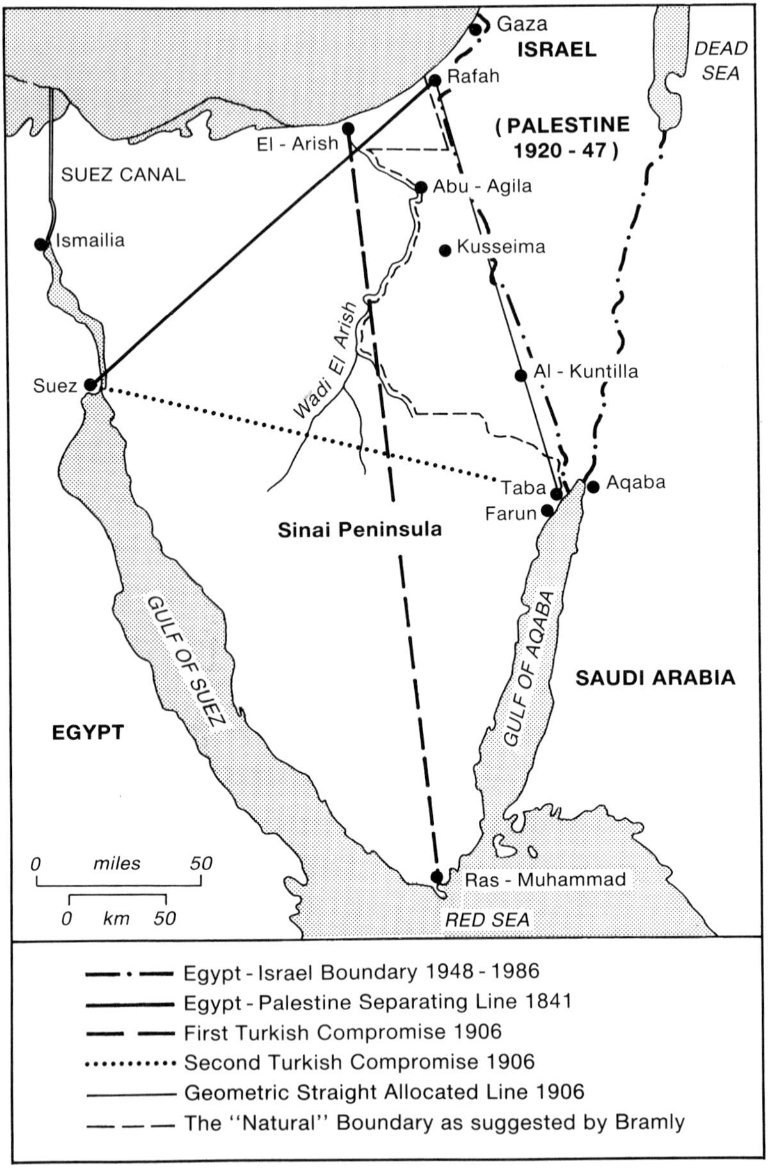

FIG 1. The Evolution of the Egypt-Palestine Boundary 1890-1914

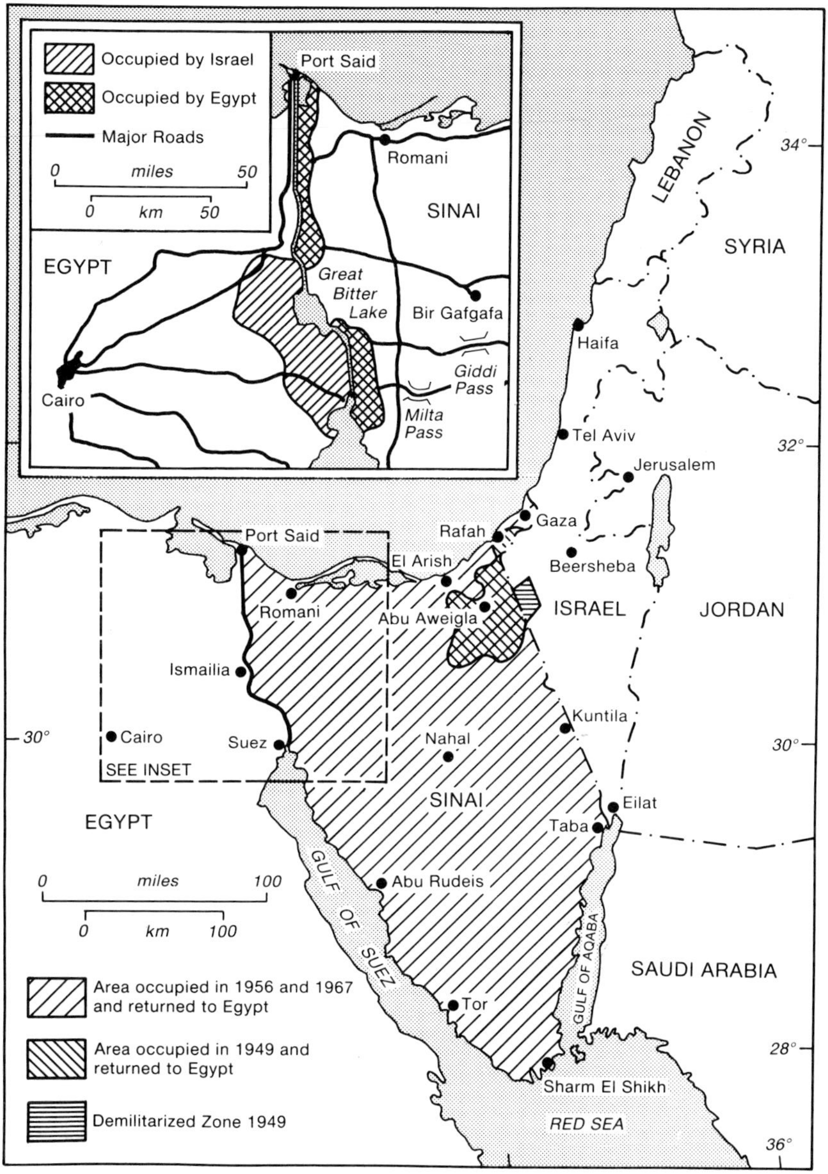

FIG 2. The Egyptian-Israeli Wars, 1949, 1956 and 1967

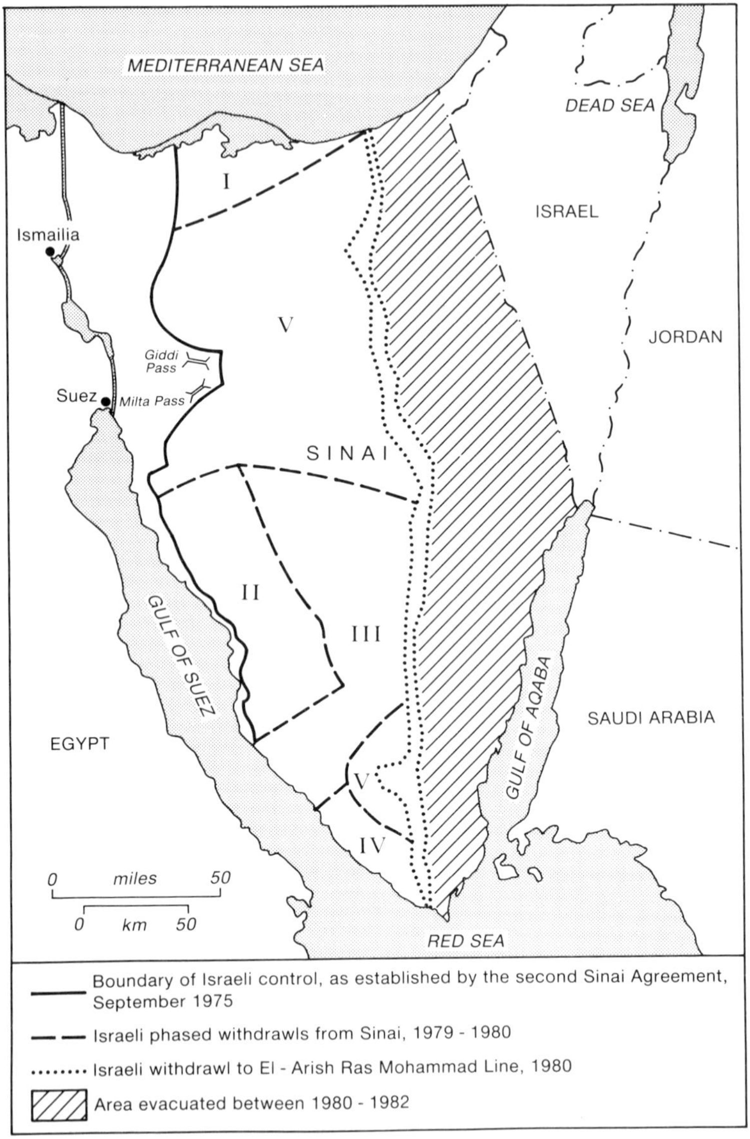

FIG 3. Phases in the Israeli withdrawal from Sinai 1979–1982

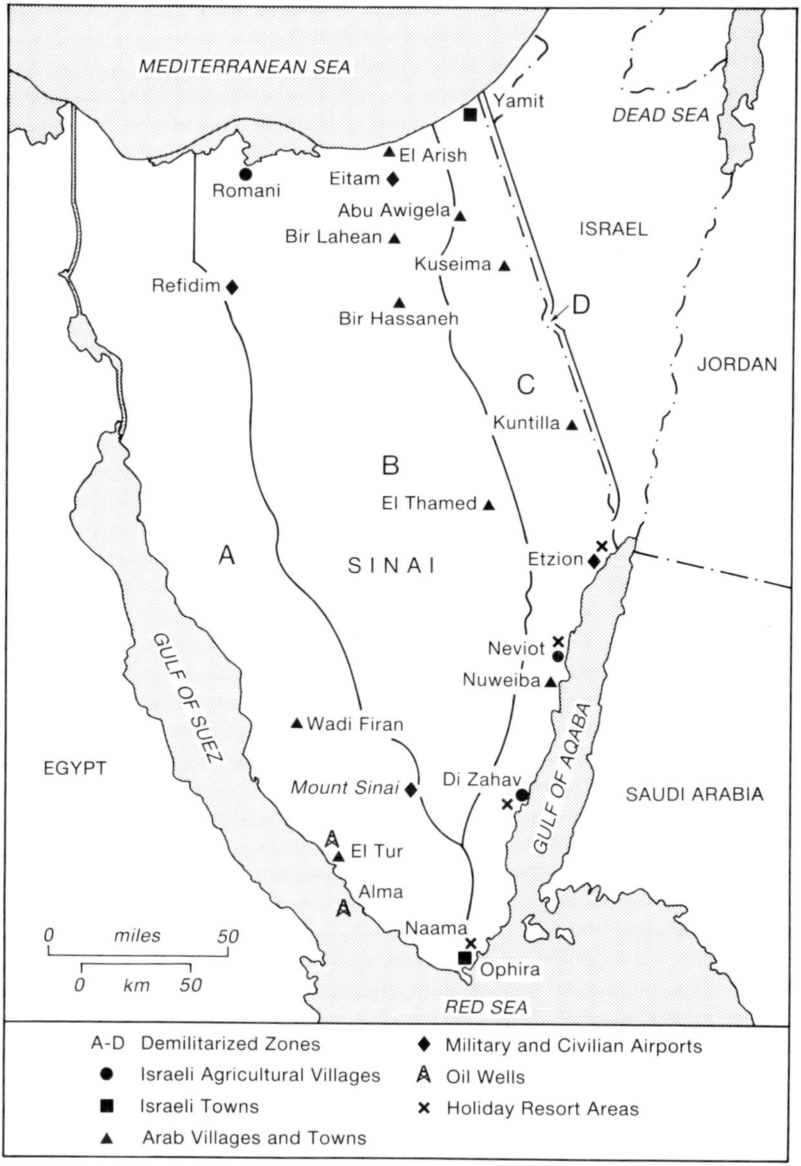

FIG 4. Israeli withdrawal from and security arrangements for the Sinai

4 POLITICAL DIVISION AND PEACEKEEPING IN CYPRUS

Carl Grundy-Warr

4.1 Introduction

> "The Cyprus Problem(s) must not be viewed simply as "a problem of Greeks and Turks". Nor is it true to say "that but for the influence of Britain and later the U.S.A. the Greek and Turkish Cypriots would to this day be living in harmony". (Loizos, 1976; 14)

> "Clearly, both communities were under the control of outside factors over which they had very little influence; there is every indication that, divorced from the wider political argument, harmony could have prevailed". (Drury, 1977; 6)

> "...ultimate responsibility for a return to normal conditions to Cyprus must, obviously, rest primarily with the authorities and people of Cyprus themselves, since normality can come only as a result of a determination by the communities ... to lay down their arms and seek to live again in peace". (Report by U.N. Secretary-General, 29 April 1964, para.2).

> "Our relationship with the Greeks was always like a cracked glass. It was shattered in 1974 and it cannot be glued back together again". (Turkish Cypriot Shopkeeper quoted in Oberling, 1982; 224).

These four quotations are selected to illustrate the complexity and paradoxical nature of the multifaceted Cyprus problem(s). Cyprus is a small, non-aligned island with a geopolitical significance far greater than its physical proportions. It is beyond the scope of this paper to explain the underlying processes behind the socio-political divisions that exist in Cyprus, and to show how the follies and phobias of Cypriots have been manipulated and aggravated by external factors/forces. Suffice it to say here, that Cyprus is enmeshed in the historical rivalry between Greece and Turkey, which continues to bedevil NATO unity, and compounds America's difficulties in maintaining military bases in both countries. Cyprus itself, with its British sovereign base areas, and its communications facility in the Troodos mountains, is an important monitoring post, for Soviet activities in the Middle East and Mediterranean. Since the collapse of the U.S. strategy in the Middle East, following the Iranian Revolution in 1979, Washington has increasingly viewed Turkey as "the new front line of the Middle East" (South, September 1986; 26), and the U.S. is helping to tie Cyprus into that "front line" by developing new facilities of probable strategic implications in the Turkish north, including a new airport. Furthermore, Cyprus is directly affected by the troubles of neighbouring states - as a refuge for Lebanese refugees and Beiruti businesses, and as a holiday resort and business centre for Arabs, Israelis and Europeans. Thus the position of Cyprus in the eastern Mediterranean has simultaneously elevated the position of Cyprus in world politics and added to its internal political difficulties, making the likelihood of a peaceful settlement between Turkish and Greek Cypriots a remote possibility, **(see Fig. 4.1).**

When looking at the political geography of Cyprus it is important to always keep in mind the fears, hardships, and psychological scars caused by political divisions between the two main communities. Every single Cypriot either has a refugee in his/her family or as a neighbour or friend. All Cypriots, whether Greek or Turkish, have had their lives dramatically altered twice since the island's independence from British colonial rule in August 1960. On both occasions the map of Cyprus was fundamentally transformed by intercommunal conflict and external interference. It is difficult for any outsider (like the author) to contemplate the bloodshed, human misery, and disruption to day-to-day existence endured by many Cypriots during the 1960s and in 1974. As Christopher Hitchens (1984) discovered.

> "It is in conversation with Cypriots themselves, however, that the even more serious wounds inflicted on Cyprus become apparent. The most casual inquiry - such as, 'Where do you come from'? or 'Do you have a family'? - can be enough to induce a torrent of grievance or of grief".

It is not the purpose of this paper to dwell on the sufferings of the Cypriot people or to delve into the dynamics of intercommunal conflict. The central theme of this paper is to concentrate on the role of the United Nations Peacekeeping Force in Cyprus (U.N.F.I.CYP.) in relation to the political divisions within the island and the everyday life of ordinary Cypriots, **(see Fig.**

4.1).

The main argument is that U.N.F.I.CYP.'s most significant role in Cyprus has not been its soldiering role of keeping the peace, but in carrying out crucial economic and humanitarian functions at the local level. This is an aspect of U.N.F.I.CYP. that is often ignored by politicians interested only in broader political goals. Admittedly the United Nations Organisation has proved to be impotent and unsuccessful in the higher level mediation attempts of the U.N. Secretary-General to produce a settlement to end the political schisms.

4.2 Turkish Cypriot Enclaves and Patchwork Partition

At independence Cyprus had a population of 650,000 out of which Greek Cypriots outnumbered Turkish Cypriots in a ratio of 4:1. The island had 114 mixed villages, and in spite of distinct ethnic quarters in the main towns and in several villages, the two communities coexisted freely and without incident. Unfortunately, the Independence Constitution, which was the product of the London-Zurich Agreements of 1959 between three outside powers - Britain, Turkey and Greece, polarized community politics. After the collapse of the constitution, inter-communal violence flared up in Nicosia and spread into the countryside and District towns. Between December 1963 and August 1964 the political geography of the island was completely transformed. The Government of Cyprus (now a Greek Cypriot administration due to the withdrawal of Turkish Cypriot officials) ruled over a Republic riddled with holes. Turkish Cypriots had evacuated 72 mixed villages and 24 wholly Turkish Cypriot villages. United Nations figures estimated approximately 25,000 refugees, of whom 21,000 were given homes in larger or "safer" Turkish Cypriot settlements, whilst the remainder found temporary shelter in refugee camps. (U.N. Document S/6102, 12 Dec. 1964, para 45).

Territorial control suddenly became a critical issue between the two community leaderships, for the Turkish Cypriots had effectively created a rudimentary, fragmented state within a state made up of tiny enclaves scattered in different parts of the island. In total, there were 44 pockets of Turkish control, that is if the barricaded Turkish quarters of Nicosia, Larnaca, Paphos and Famagusta are included. Of the 135 Turkish Cypriot centres partially or fully occupied in August 1964, only twenty were under the direct territorial control of the Cyprus Government. (refer to Map 2) These enclaves and quarters were to become "the geographical stepping stone" to a **de facto** Turkish Cypriot state. (King 1980; 8)

Nowhere was the physical segregation of Greek and Turkish Cypriots more pronounced than in the old walled city of the capital, Nicosia, where the "Green Line" carved through the city like a cheese-wire. This line was drawn by a British Army officer on the night of 28-29 December 1963 and it was never intended to be anything more than a temporary **cordon sanitaire**, separating the armed groups of each community. In fact it has become a symbol of ethnic segregation and a division of international geopolitical significance, an unremitting obstacle to progress towards harmonizing relations between the two communities, **(see Fig. 4.2)**.

By the time U.N.F.I.CYP. became operational on 27 March 1964, relieving the beleagured British peacekeeping force, the island was already crisscrossed by barbed-wire fences, road-blocks, trenches and other fortifications. U.N.F.I.CYP's initial task was to interposition troops between the Greek National Guard and Turkish Cypriot fighters. Three concentric rings of soldiery were established around the main enclaves and along the green lines of Nicosia, Larnaca, and Pathos. In such a way the U.N. Force contributed to the pacification of Cyprus, but as Professor Coufoudakis argues, "it also froze the situation so that the Turkish Cypriot enclaves and lines of division in the cities remained intact for a decade". (Coufoudakis, 1976; 459) Nevertheless, this point of view underplays many non-military civic action duties carried out by U.N.F.I.CYP. in its efforts to contribute to a "return to normal conditions" as requested by its mandate. Such activities sought to lessen the hardships caused by artificial physical divisions and break down some of the psychological barriers between the two communities. Indirectly the Force was trying to do more than maintain a negative and unstable spatial **status quo** between two antagonistic communities, for it was trying to form a communication bridge between them, through its mediation and good offices.

The formation of separate areas of control made the Turkish Cypriot community like Macbeth - "cabin'd, cribb'd, confined, bound in to saucy doubts and fears". (Shakespeare, Macbeth, Act III, Scene IV) These introverted, claustrophic Turkish enclaves were too small with too few resources to become economically viable or self-supporting units, even at rudimentary subsistence levels. Furthermore, their segregation from the rest of the Cypriot economy reinforced the growing economic dichotomy between the two communities -

> "...for ten years during which the Cypriot economy boomed and standards of living rose visibly on all Greek sides, the Turkish element was left to stew in its own juice". (Drury, 1981; 300)

The relative isolation of the enclaved Turkish Cypriots from the island's economic growth was enhanced by the mutual mistrust displayed by the two leaderships. On the one hand, the Provisional Turkish Cypriot Administration refused to cooperate with the Greek Cypriots in any project which involved an extension of or recognition of government authority within Turkish areas of control. It promoted separate economic development projects utilizing an annual grant from Turkey. On the other hand, the Cyprus Government built new infrastructure around Turkish-controlled territory, which deprived the Turkish Cypriots of numerous basic civilian services and amenities. Until the lifting of a government economic blockade on Turkish Cypriot enclaves in March 1968, U.N.F.I.CYP. was the only link they had with the outside world.

4.3 The Partition of Cyprus and the U.N. Buffer Zone

Unfortunately, U.N.F.I.CYP. could do little to influence either the Greek or the Turkish Cypriot leaderships, and it had no power at all to influence the thinking of politicians in Greece, Turkey, and the United States. Interference by these external powers in Cypriot affairs led to the **de facto** partition of Cyprus into two virtually mono-ethnic zones. On 15 July 1974, there was an Athens Junta-backed **coup d'etat** by the National Guard against Archbishop Makarios, installing Nicos Sampson, a former E.O.K.A. anti-Turk terrorist as leader of the 'new regime of national salvation'. In response, Ankara ordered the military intervention of Turkish troops on 20 July, who eventually succeeded in occupying 37% of the island before an island-wide cease-fire was called on 16 August. One immediate result of this military division of Cyprus was a massive population exchange. Between July 1974 and December 1975, about 185,000 Greeks moved southwards and 45,000 Turkish Cypriots moved northwards. The north of the island was renamed the 'Turkish Federated State of Cyprus' on 13 February 1975, a **de facto** micro-state without international recognition.

The creation of the so-called 'Attila Line' or Turkish cease-fire line led to sudden and dramatic changes in U.N.F.I.CYP. deployment. Observation posts and patrols were established in six contingent sectors along "the general line of confrontation", whereas previously U.N. soldiers were stationed at strategic points throughout the island's six Districts, with permanent observation posts concentrated around the Turkish enclaves. Between the forward defence lines of each side U.N.F.I.CYP. controls a 'neutral' strip of land extending the full length of the Turkish F.D.L., (forward defence line) for 180 kilometres, varying in width from seven kilometres at its widest point to just 20 metres at its narrowest within the city walls of Nicosia. In total, the U.N.F.I.CYP. Buffer Zone accounts for approximately three per cent of the total land area of Cyprus and almost ten per cent of agricultural land, including a tiny proportion of irrigated land near Morphou. The Buffer Zone cuts across the southern extension of the Morphou citrus groves, slices through the heart of the cereal growing Mesaoria plain, dissects the capital city, truncates many pre-August 1974 routeways, and runs over important river basins, streams and underground water resources, **(see Fig. 4.3).**

Within the walled city of Nicosia the thin blue line of U.N. soldiers is sandwiched in between the Turkish Forces and the National Guardsmen along the Green Line. Although U.N.F.I.CYP. has proved successful in its efforts to maintain the spatial **status quo** since 1974, its most significant contribution to Cypriot life has been in transforming otherwise barren land "between the lines" into farmland once again. Unproductive no-man's-land close to the F.D.L.s of each side is being cleared of minefields, and farmers are being encouraged to recultivate agricultural land in the Buffer Zone. A Farming Security Line has been carefully drawn and demarcated by U.N.F.I.CYP. through the Buffer Zone for the mutual safety of civilians working on either side. This security line has developed into perhaps the most politically and militarily sensitive line in the island. The Peacekeeping Force referees all activity within the Buffer Zone and is able to permit or refuse any farmer to work in that area. Permission to cultivate is given "without prejudice to any eventual political settlement" under two principal guidelines:-

(i) Security - cultivation should not be a threat to the legitimate security interests of either side:

(ii) Ownership - U.N.F.I.CYP. has tried to ensure that the land is cultivated by actual owners or their employees.

Documented proof of ownership must be submitted to U.N.F.I.CYP. before a farmer is allowed to cultivate fields in the Buffer Zone. At all times security interests are given priority over ownership rights, and U.N.F.I.CYP. does not allow cultivation where it is considered to be in a threatening position to either side, or where farmers lives are at risk.

At present the farming security line forms an effective ethnic interface within the only zone in Cyprus where Greek and Turkish Cypriots could still live and work side by side. It should be stressed that the farming security line is not meant to be a static or permanent dividing line, and that U.N.F.I.CYP. are willing to make adjustments to the line in order to extend economic activity, as long as it is safe to do so. Hitherto, U.N.F.I.CYP. has succeeded in extending Greek Cypriot cultivation south of the farming security line, while most of the land north of the line is scrub or rough grazing-land for goats. This is partly because the Turkish Army control and restrict the activity of Turkish Cypriots in front of their trenches and sentry posts, and partly due to the fact that the Turks gained more land per head with the **de facto** division and have little need to utilize land within the Buffer Zone. Furthermore, U.N.F.I.CYP. does not allow the Turkish Cypriots to cultivate Greek owned land north of the farming security line except in a few restricted areas.

Since July 1974, U.N.F.I.CYP.'s activities in the north of the island have been restricted to routine humanitarian visits to Greek Cypriots and a small Maronite community remaining there. There are now only 800 or so Greek Cypriots residing in the north, most of whom live in the tobacco-growing village of Dipkarpaz (formerly Greek Rizokarpaso) at the tip of the Karpass peninsula or 'pan-handle' of Cyprus. U.N. civilian police patrols pay frequent visits to these "minority groups" in the now "Turkified" north, and they maintain regular contacts with the police and village leaders of both sides. The main task of the U.N. civilian police, however, is to ensure the protection of civilians (estimated to be over 9,000) and property within the Buffer Zone, and to prevent civilians crossing into areas on the "wrong" side of the farming security line. Within this U.N. demilitarized zone there are villages such as Pyla where life goes on much as it did for decades before political and physical partition. Functioning Orthodox churches and mosques stand side by side, and the Turkish and Greek coffee houses still cater for their regular all male clienteles. The Turkish children of Pyla are perhaps the last of their nationality who will speak Greek, particularly with the continued "Turkification" of northern Cyprus. Children in each community go to their separate schools, although a Greek kindergarten in Pyla still accepts Turkish pupils.

Under the auspices of the U.N. "blue berets", and with the continuation of their economic and humanitarian duties, the Buffer Zone could eventually become an area of intercommunal coexistence and activity, although this would obviously require considerable guarantees and compromises from both sides. In the meantime, U.N.F.I.CYP. are helping to close the physical gap between the two communities by promoting the productive use of the snake-like stretch of land which separates them.

Political deadlock between the Greek and Turkish community leaderships continues to thwart various U.N. peace settlement proposals. Successive plans presented by the U.N. Secretary-General, Senor Javier Perez de Cuellar have been rejected either by one side or the other. A 1985 summit meeting in New York between the community leaders under United Nations auspices broke down without agreement on a number of fundamental issues. President Kyprianou, the Greek Cypriot leader, was not prepared to enter an agreement on the future of the island, which would effectively legitimise the Turkish Cypriot administration, before plans were made for a withdrawal of Turkish mainland forces from northern Cyprus. Since **de facto** partition little progress has been made in intercommunal discussions and debates about possible federal solutions or proposed territorial concessions to the Greeks. Other crucial sticking points in intercommunal negotiations include, differing concepts of "bizonality"; the problem of equal political status for the communities (the Greeks form 80% of the island's population); and the issue of freedom of movement and settlement throughout the island for Greek Cypriots, and particularly for the refugees of 1974. Another problem is that of guarantors of a settlement. The Turkish Cypriots want Turkey as a guarantor power, whereas the Greek Cypriots would prefer guarantors with no direct interest in the island. Senior Greek-Cypriot officials have been working for an international conference on the Cyprus Problem(s), but this idea is opposed by the U.S. on the grounds that such a conference would give Moscow a greater role in settling the Cyprus problem.

It also wrong to assume or to treat an ethnic group as if it was a single thing with one mind. Both Cypriot communities have complex class and social structures with a range of political opinion, from communists, and revolutionary socialists, through laissez-faire pro-capitalist liberals, to traditionalist conservatives, and extreme rightists of fascist persuasions. (Loizos, 1976; 14) Thus on the Greek Cypriot side President Kyprianou has frequently been criticised for compromising too much or on other occasions, not enough, with the Turkish Cypriots. His main critics come from the two main opposition parties - Akel, the communist party, which has maintained some contact with fellow leftists on the Turkish side and urges more dialogue, and Glavkos Clerides's Democratic Rally Party. In the Turkish Cypriot part of the island all the former Greek village names have been changed into Turkish - Kyrenia is now Gime, Morphou is Guzelyurt, Famagusta is Gayi Magosa. As well as 20-30,000 soldiers there are an additional 40,000 or so settlers from the mainland - 'Turkiyeli', which may be part of a deliberate policy to alter the demographic basis of the island. Many of these Turkiyeli come from relatively backward parts of Anatolia and the Black Sea coast, have different modes of dress, are less 'European' in their behaviour and values, and adhere more fervently to traditional Muslim rights than do Turkish Cypriots, all of which has caused some feelings of tension between the "islanders" and the mainland settlers. Meanwhile Turgut Ozal's much publicised visit to northern Cyprus has strengthened existing political ties between Ankara and the Denktas leadership, although the Turkish Cypriot business community are unenthusiastic about Ozal's promises to make the area a vast offshore zone for the benefit of Turkish entrepreneurs.

Whilst the politicians on both sides argue their respective (and often diametrically opposed) cases, the physical dividing line has hardened into an effective boundary between two micro-states.

On the 15th November 1983, Rauf Denktas, the self-styled President of the Turkish north, proclaimed the independence of the 'Turkish Republic of Northern Cyprus'. This act consolidated a decade of **de facto** partition and two decades of separation from the Republic of Cyprus, or as Denktas prefers to put it:-

> "We have not seceded from the Republic of Cyprus, we have been thrown out of it and we have had to put the roof of statehood over our heads", (South, January 1984; 21)

The north still lacks international recognition and is totally dependent on Turkey for military, financial and economic support, but as time goes on Denktas' 'Lilliputian state' (as it has been described by some Greek Cypriot leaders) could well receive international recognition, **(see cartoon in Fig. 4.4).** In the meantime leaders of both communities continue to manipulate the nationalist passions that have helped split the island, inflaming old fears of 'Enosis' (Union with Greece) on the one side and **taksim** (partition) on the other. As Keith Kyle wrote in the 1984 Minority Rights Group Report on Cyprus,

> "Phobias will have to be overcome on both sides...time is running out...as years pass and new generations grow up fewer people will have any personal experience of living with members of the other Cypriot community..."

Unfortunately, the interests and hopes of ordinary Cypriots on both sides have always been secondary to the strategic interests of outside powers. The people of Cyprus have paid a high price for their differences - losing lives, land and homes in past intercommunal struggles. At all times their intercommunal differences have been used or manipulated by various outside powers. The political and military manoeuverings of such powers frequently disregard the people who are most directly affected by their actions/decisions. Meanwhile the bitter memories of past conflicts persist between the Greek and Turkish Cypriot communities, and generations of children will continue to grow up in mutual ignorance of the other side. Politicians on both sides continue their posturing and continue to spout 'hot air' about a political settlement but make no progress towards one. The dividing line is thus a more solid barrier to communications between Greek and Turkish Cypriots than many ratified international boundaries, with the United Nations Force continuing its dual role as buffer force and communication bridge between the two communities.

It is fitting to end with the words of Rina Catselli a refugee from Kyrenia (although the words are appropriate for refugees everywhere):-

> "You try to forget, but the dreams of a refugee are merciless. They take you to places you knew and give the delusion that everything continues to be as it was before. Then comes the painful awakening". (Catselli, 1979; 47)

4.4 Acknowledgement

I should like to thank Margaret Bell for her assistance in preparing this paper.

References

Catselli, R. (1979) **Refuge in My Homeland**, Nicosia: "Avghi" Printing Co.

Coufoudakis, V. (1976) 'United Nations peacekeeping and peacemaking and the Cyprus Question', in **Western Political Quarterly** (USA), Vol. 29, no. 3, Sept. 1976, pp 457-473.

Drury, M.P. (1981), 'The Political Geography of Cyprus', in Clarke, J.I. & Bowen-Jones, H, (editors), **Change and Development in the Middle East**, Methuen, London.

Drury, M.P. (1977), **Western Cyprus : Two decades of population upheaval, 1956-1976.** Paper presented to the Institute of British Geographers, Annual Conference, 7 January, 1977.

Hitchens, C. (1984), **Cyprus**, Quartet, London.

Kamm, H. (1986), 'In the U.N. Buffer Zone, a Cypriot village evokes old ethnic harmony', **International Herald Tribune,** 11/7/86

King, R. (1980), 'Cyprus since 1974 : Economic and Demographic Change'. Paper presented to the Geography Sections, **Fifth Mediterranean Conference,** Bar-Ilan University, Tel Aviv, 5-7 August 1980.

Kyle, K. (1984), **Minority Rights Group Report : Cyprus**, Report No. 30, Minority Rights Report

Loizos, P. (1976), quoted in **Minority Rights Group Report : Cyprus**, Report No. 30, Minority Rights Report.

Oberling, P. (1982), **The Road to Bellapais**, Columbia University Press, New York.

Patrick, R.A. (1976), **Political Geography and the Cyprus Conflict, 1963-1971,** University of Waterloo Press, Ontario, Canada.

South, Third World Magazine, January 1984, p. 21. Interview with Rauf Denktas.

South, Third World Magazine, September 1986, pp. 26-27, 'Divided We Fall' report on Cyprus.

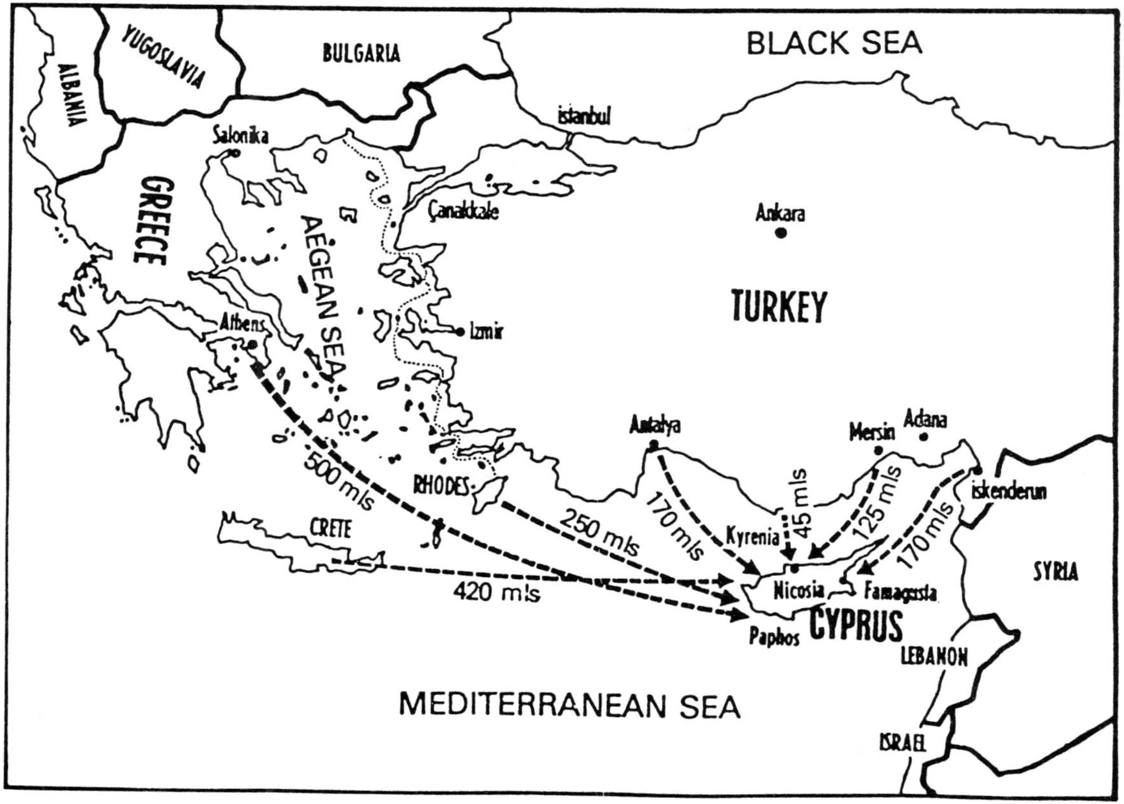

FIG 1. Geopolitical situation of Cyprus with regard to Greece and Turkey

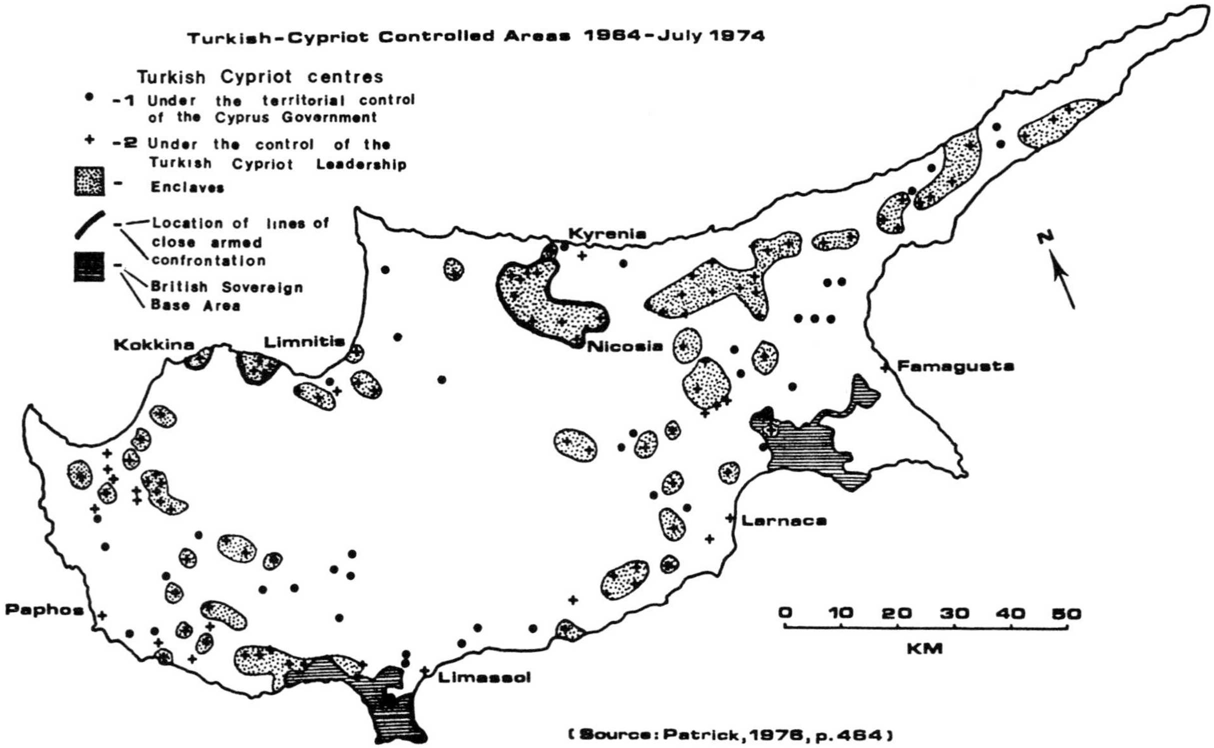

FIG 2. Turkish Cypriot controlled areas, 1964-July 1974

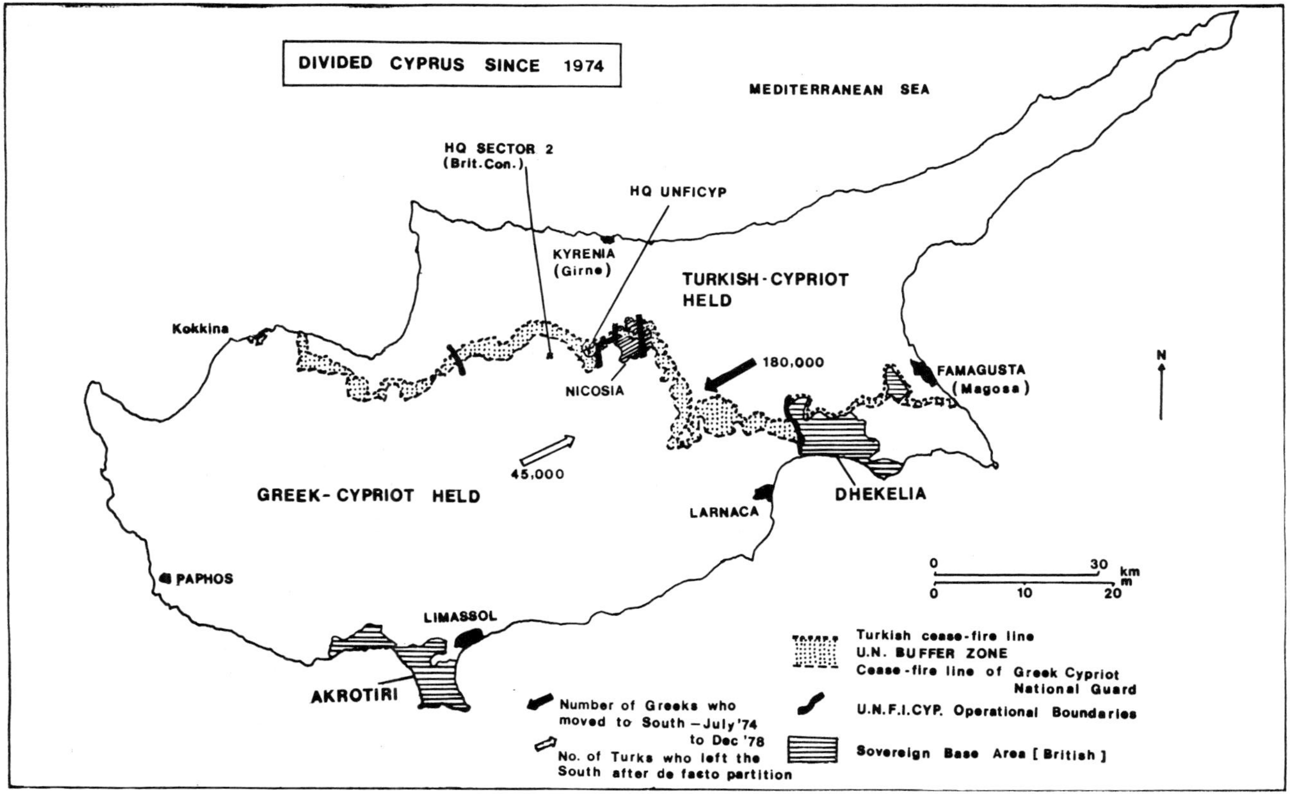

FIG 3. Divided Cyprus since 1974 detailingU.N. Buffer Zone

FIG 4.

Greek Cypriot cartoon depicting Rauf Denktas, the self-proclaimed President of the Turkish Republic of Northern Cyprus, as the villain of the current political situation by cutting off Northern Cyprus from the rest of the island.

On 15 November 1983, Denktas declared U.D.I.

5. WATER RESOURCES & BOUNDARIES IN THE MIDDLE EAST

Ewan W. Anderson

5.1 Introduction

During the middle and latter part of this century, oil has been the liquid dominating resource geopolitics in the Middle East. However, increasingly it is being realised that water is now the fundamental political weapon. Indeed, it can be confidently prophesied that despite the inevitable growth in Western dependence upon Persian/Arabian Gulf oil towards the end of this century, the politics of the region will be shaped more by water than by oil (Perera J, 1981).

Throughout most of the Middle East, rainfall averages under 100mm per annum, although around the desert fringes 200mm may be recorded and in the higher areas up to 400mm. For agriculture this is crucial since cultivation requires at lease 400mm, while areas with less than 250mm can only be used for rough grazing. The major exceptions, in which falls of 1000mm a year or more are recorded are the higher mountains of the Lebanon and the Maghreb, together with restricted areas of Turkey and Iran. Thus, except where relief intervenes, the Middle East is in general unsuitable for agriculture. However, with comparatively high rates of population growth throughout the area, the need for increased economic development, particularly in agriculture, is obvious. Indeed, food is the most strategic of all supplies and to move nearer self-sufficiency is the aim of all the countries in the area. Therefore, it is vital to utilize all possibile water supplies and to develop potential new sources. This necessity is clearly recognized in the high priority given to water policy by all the governments and throughout the Middle East there is a high level of investment in water exploration, hydraulic engineering and the development of alternative supplies.

It is however, a complex problem. For example, the lower the annual total, the less reliable the rainfall becomes, so that several years of drought may be followed by the incidence of catastrophic floods. While every effort has to be made to extract water, the prevention of massive losses through runoff to the sea must also be kept in mind. A system of recharge dams, constructed to make the greatest use of any flow, may retain no water at all for three years and in the fourth be totally overwhelmed by the floods. A further crucial problem concerns increasing levels of salinity in the soil. High evaporation rates, together with the meticulous use and re-use of water, results in enhancemant of the dissolved mineral content. Furthermore, the rapid development of irrigation depletes the ground water, lowering the seaward pressure gradient and allowing the incursion of saline water from the sea. Thus, the hydrological relationships are subtle and the mere infusion of finance, even on a vast scale, is unlikely to solve all the difficulties.

Many other problems, hydrological, economic and social, can be identified in analyzing the current and growing crisis in Middle Eastern water supplies. However, there are two factors, both essentially political, which exercise an overriding potential influence. Firstly, with regard to the natural supply, the boundaries of water sources, both surface and sub-surface, frequently do not coincide with political boundaries. With surface flow there is an obvious basis for competition, if not actual conflict. However, less clear cut, but increasingly significant is the fact that sub-surface abstraction on one side of a boundary may seriously affect supplies on the other. Not only are the few main surface catchments divided by political frontiers, but in many cases, the aquifers are similarly dissected.

In an area with such a wide range of potential conflicts, both regional and local, including many border disputes, water supplies may also represent a strategic weakness. By the nature of their infrastructure, water supplies may be comparatively easily disrupted, but this situation is greatly enhanced when these supplies depend essentially on one source. The growing dependence upon desalination in the Middle East has in fact resulted in the case of many countries in the reliance upon a very few major installations. These could be relatively easily damaged with virtually immediate and far-reaching results.

Using population projections, the full extent of the Middle Eastern water crisis can be appreciated. The table lists statistics of water availability, both surface and ground, per capita in 1000 cubic metres per annum:

5.2 Natural Sources

All the perennial rivers of the Middle East, with the exception of the Nile, are to the north of latitude 30°N, while even beyond that parallel there are considerable areas with only ephemeral

TABLE 5.1 : PER CAPITA WATER AVAILIBILITY (in 000s m3)

	1971	2000	Population Increase %
Algeria	2.20	1.00	111
Arabian Peninsula States	0.70	0.30	106
Cyprus	0.06	0.05	22
Egypt	0.10	0.05	111
Iran	6.00	2.50	145
Iraq	3.60	1.30	173
Libya	3.70	1.20	198
Morocco	2.10	0.90	132
Sudan	4.00	1.90	107
Syria	3.00	1.00	165
Tunisia	0.90	0.40	126
Turkey	4.90	2.3	118

Source : Barney, 1980

Taking the accepted minimum as 1,000 m3 per capita, per annum, it can be seen that there is a shortfall, in many cases marked, in most countries.

surface flow. The opportunities for barrage construction are thus limited and, on perennial rivers multiple damming is becoming the norm. Such installations and particularly major multi-purpose dams exercise a great influence on developments downstream and this can be very significant when a river crosses a political frontier. For example, the Ataturk Dam on the Turkish section of the Euphrates, to be completed by 1991, with a capacity to irrigate approximately 750000 hectares, will clearly affect developments in Syria and Iraq. Despite the fact that there are at least 35 major dam schemes projected for the permanent rivers of the region, the most marked tendency is towards constructions on seasonal water courses. In some cases, these may impound sufficient water to form a permanent reservoir, but in most they are designed to check losses by increasing recharge into the aquifers. Saudi Arabia alone has plans for 60 such dams, to be completed by 1990. Two major constructions of this kind, each over 20km in length have already been completed on the Batinah plain of Oman.

For millennia, throughout the region ground water has been, in almost all cases, the major source of supply. Commonly, it has been obtained from springs and wells, using, in the latter case, lifting devices. However, in Iran and in parts of the Arabian Peninsula, notably Oman, subterranean canals with associated surface distribution systems, known as qanats or aflaj, have been of great importance (Wilkinson, 1977). With the dramatic increases in water requirements over the past few decades, there has been a strain on all these sources. This applies particularly of course to those countries with the greatest reliance upon ground water, such as Israel, Libya and the Gulf States. Indeed, almost half of the Oman Ministry of Agriculture's budget allocation for water in the next five year plan has been set aside for the maintenance and repair of aflaj (Bodgener, 1984).

The other measure with regard to ground water is the use of deep "fossil water" aquifers, although these are considered by many experts to constitute a non-renewable resource. The most spectacular example at the present time is the investment by Libya of over U.S. $3,300 million in the "Great Man-made River Project" to irrigate some 180000 hectares. When complete, the "river" will stretch from Kufra in the southern Fezzan to the coast and will facilitate agricultural and industrial development around a large part of the shoreline of the Gulf of Sirte (Odone, 1984). However, in the medium term at least, such a project can constitute only a temporary solution.

With both surface and ground water sources under stress throughout the region, there is a great urgency attached to the search for alternatives. Chief among these by far has been desalination and more money has been spent on installations in the Middle East than in any other part of the world. Indeed, the region possesses over 35% of the world's desalination plants and over 65% of the total desalting capacity. There has already been extensive research to cut costs of water produced by this method and even more dramatic developments can be expected when the new prestigious Qatar Solar Energy Research Station has been completed. At present the most promising method and one which can be replicated on a small scale is that of the saline solar pond in which the sun's radiated energy can be collected and stored (Keen, 1985).

Another alternative, but one limited to irrigation, particularly, is reprocessed sewage. Water from this source is already in use in many countries such as Jordan, Saudi Arabia and Qatar, while in Kuwait there are plans to irrigate up to 16000 hectares from this source. Other possible solutions, often mentioned but rarely fully accurately costed, involve the regular import of water. The possibility of towing icebergs from the Antarctic seems feasible, but such practicalities as the control of melting rates and the distribution of the resulting water have not been properly addressed. A further suggestion and one already implemented to a certain extent is for the importation of water by products tanker. Chief among the countries promulgating such an idea have been France, Japan and the United Kingdom, each with well placed regions of surplus and a well established shipping infrastructure. If the amounts concerned were substantial, the logistics would be very complicated while, for any consumer country to place long term reliance upon a foreign source for such a vital commodity as water seems highly imprudent.

5.3 Water Geopolitics

While water scarcity itself will for long remain the key underlying problem, the lack of accord between resource boundaries and political frontiers is more likely to produce overt confrontation. Upstream abstraction will affect both the quantity and quality of water available for down-stream use. The most obvious geopolitical flashpoints are those involving surface water in the two major drainage basins of the region, the Nile and the Tigris-Euphrates, but also perhaps in more spectacular form, in that of the Jordan **(see Fig. 5.1).** Two other rivers of significance in this context are the Orontes (which flows through Lebanon, Syria and Turkey) and the Litani, the whole course of which is in Lebanon. Elsewhere in the region, there is surface flow in Turkey, Iran and the Maghreb, but there are no obvious sources of conflict. For example, apart from the major catchments of the Tigris and Euphrates, the catchments of Asiatic Turkey, whether flowing to the Black Sea or the Mediterranean are basically self-contained. Iran includes the Karun together with some of the more minor head waters of the Tigris, but these are mainly in restricted valleys and dam building, for whatever reason, seems unlikely.

5.3.1 The Basin of the Nile

Since it flows through so many different climatic regions no river has such a complex hydrological regime as the Nile. Furthermore, from its confluence with the Atbara to the Mediterranean, a distance of some 1800km, there are no perennial tributaries. With a rapidly growing population of approximately 47 million, virtually settled in the Nile valley, the need to increase the agricultural area of Egypt is vital. Therefore water use must increase and the sole supply of importance is the Nile itself.

Since in Egypt water requirements underlie every facet of life, it is hardly surprising that estimates for supply and demand vary wildly. Even the most optimistic estimate yields a very small surplus. These calculations depend principally on four assumptions (Whittington and Haynes, 1985):

i) Crop water requirements per hectare in the Old and New Lands (including conveyance losses in the irrigation distribution system.
ii) Extent of agricultural expansion through desert reclamation programme.
iii) Completion schedule of upper Nile water conservation projects.
iv) Extent of drainage water re-use in Egypt.

By 1990, the Egyptian Water Master Plan foresees a surplus of over 8000 million cubic metres (mcm) per year. In contrast, the Waterbury assessment indicates a deficit of some 4000mcm annually. Whichever is more realistic, the fact remains that to expand its cultivated area, Egypt has three basic options:

i) To increase the efficiency of the irrigation system and improve farm water management practices.
ii) To utilize more efficient irrigation and drainage technologies.
iii) To increase the re-use of drainage water.

Statistics for Sudan are even less reliable and more contradictory and by 1990 deficits almost as high as 14000mcm per year have been forecast. However, the worst forecasts result from over-optimistic plans to make Sudan a major world agricultural producer.

The average annual discharge of the Nile as it enters Egypt is some 85 billion cubic metres (bcm), 25bcm being derived from the White Nile with its head waters in Sudan and Uganda and the remaining 60bcm form the Blue Nile (50bcm) and the Atbara (10bcm), both rising in Ethiopia. In 1929 the first extensive Nile water agreement was reached, Egypt receiving 48bcm and Sudan 4bcm per annum, leaving approximately one-third of the Nile waters to pass unused to the Mediterranean (Whittington and Haynes, 1985). The proposal for the Aswan High Dam, completed between 1959 and 1971, necessitated a new allocation - the Agreement for the Full Utilization of the Nile Waters which was signed in 1959. This followed a period of increasing tension and, in 1958, military confrontation between the two countries. The dam, completed by 1971, yielded a further 32bcm, although 10bcm is lost through evaporation from Lake Nasser. Of the remaining 22bcm, under the 1959 agreement Egypt received 7.5bcm and Sudan 14.5bcm. To date there has been no conflict over this allocation. In fact, there have been several joint water projects, the biggest being the construction of the Jonglei Canal through the Sudd marshes which, when completed, should recover some 4.7mcm of water annually (Charnock, 1984). Currently, as a result of the war in southern Sudan, the programme has lapsed.

On the other hand, future developments are far more controversial and in particular those involving the African states will greatly affect the geopolitics of the basin. The fact that the Nile head waters are controlled by such states as Ethiopia, Uganda and Zaire exercises a strong influence on Egypt and, in particular, the advent of the present regime in Ethiopia had engendered fears of a possible politically motivated dam building programme. In contrast, the African states are concerned about and highly critical of potential Egyptian plans to supply Nile water to the Sinai and even the Negev Desert of Israel and also of the possible Sudanese plan to provide 20mcm a year by pipeline to Saudi Arabia.

5.3.2 The Basin of the Tigris-Euphrates

Although it is the only major basin in the region with a marked surplus of water, the Tigris-Euphrates river system still faces many current and latent geopolitical problems. None of the riparian states concerned, Turkey, Syria or Iraq, is actually facing a water shortage, but problems of management, apportionment and development planning have all led to controversy. All three have plans for new large scale irrigation and hydro-electric power generation projects and tensions are bound to rise, particularly in Iraq which would of course feel the full effect of any upstream changes.

As both rivers are subject to major flow fluctuations, seasonally and from year to year, one function of upstream dam building, water abstraction, may lead to disputes, but another, that of discharge control, is beneficial. The mean annual disharge of the Euphrates is approximately 32bcm, that of the Tigris 42bcm and that of the Karun 20bcm. All three carry comparatively large amounts of sediment, often excessive in the case of the Tigris and the Karun, and all suffer large losses from evaporation during the summer months. Since the three riparian states are aligned geographically in linear fashion, water quality is of great significance. Water draining back after irrigation will tend to have an enhanced salinity, particularly with the high rates of evaporation in the region.

Iraq was the first of the three states to begin major construction with the Hindiya barrage, completed in 1913 and today it is estimated that almost 50% of Iraq's agricultural area is under irrigation. Iraq is considered to be the one Middle Eastern country to be self-sufficient in agriculture based on irrigation (Allan, 1985). Syria is also heavily dependent upon agriculture and after the initial exploitation of the Orontes, attention turned to the Euphrates. The main construction was the Ath-Thawrah project and it was expected that the irrigated region resulting would total anything from 200,000 to 500,000 hectares, but so far this has not been achieved. It now seems clear that all the Syrian projects are behind schedule.

Turkey has potential schemes for both the Euphrates and the Tigris, but so far the accent has been heavily upon the former. The first was the Keban Dam, finished in 1973, and as the impounded lake was filling at the same time as that of the Ath-Thawrah Dam in Syria, the temporary effect on flow was significant. Of the other major schemes, either planned or under construction, by far the most important is the Ataturk Dam which will require some 10bcm of water annually and will, it is hoped, irrigate as much as 700,000 additional hectares. If this abstraction is added to the 7bcm projected from the Syrian plans, and allowance is made for the additional evaporation resulting from the large lake surfaces, the amount of water in the Euphrates entering Iraq would be reduced from the present figure of 30bcm to 11bcm per annum. This last statistic is particularly crucial since Iraq claims that its future minimum requirements will be 13bcm. Furthermore, it considers that, as a result of the irrigation projects at present operational, there has been a marked deterioration in the quality of Euphrates water. So far, there have been no political agreements, but in 1984 Syria called for the establishment of a multi-national Euphrates River Authority and for a joint meeting to discuss riparian rights (Drysdale and Blake, 1985).

The only documented crisis related to water in the basin occurred between Syria and Iraq in 1974, when the effect of new Syrian and Turkish dams reduced flow in the Euphrates to approximately 25% or normal. Various threats were issued, including the bombing of the Ath-Thawrah Dam and troops were amassed along the frontier. However, following the intervention of Saudi Arabia, Syria agreed to release additional water from the Ath-Thawrah Dam in June 1975 and normal relations were restored.

The Shatt al-'Arab, produced by the confluence of the Euphrates and the Tigris, and receiving additional flow from the Karun, presents rather different problems. As the lower part of the river course provides the border between Iran and Iraq, this concerns frontier rather than water issues. Indeed, during the long history of the dispute, few practical problems concerning water have ever arisen. However, since defensive moats have been constructed by Iraq to protect the city of Basra, the picture could change if these were to dry out as a result of upstream abstraction.

5.3.3. The Basin of the Jordan

However, the most intractable problems are clearly in the basin of the Jordan River, divided between Lebanon, Syria, Jordan and Israel. Not only is the Jordan, by any standards, a small river, with a total discharge equivalent to about 2% of the annual flow of the Nile, or 7% of the Euphrates in Syria, but with the exception of Lebanon, the countries involved all face major water shortages. The most important source of the upper Jordan is the Dan Spring which contributes approximately 50% of the discharge, the other major contribution being from the Yarmuk River (Naff and Matson, 1984). The upper Jordan itself accounts for approximately 40% of Israel's water budget and it is significant that only 3% of the area of the Jordan basin lies within the boundaries of pre-1967 Israel.

In Israel the main development, the National Water Carrier, completed in 1964, and therefore lying entirely within Israel's pre-1967 boundaries, transports water from Lake Tiberias, along the coastal plain to the Negev Desert. There are contributions from other sources to the flow which averages 320mcm per year. The other major project was the Mediterranean-Dead Sea Canal, planned to have an annual flow of 725mcm, but plans for this have now been shelved. Other options for Israel include the greater use of ground water and there is evidence that exploration in the south has been successful. Desalination is considered less cost-effective than recycling from sewage, but the overall cost of artificial sources is high and it is clear that Israel, within its present boundaries, will always have major water problems.

At an early stage, Jordan was involved in co-operative efforts with Syria in initiating the Great Yarmuk Project. A major part of this, the Upper East Ghor Canal, was completed in 1964 and further expansions have occurred subsequently, although it has not reached, as originally planned, the Dead Sea. Jordanian plans to construct the Maqarin Dam on the Yarmuk to help alleviate its problems, have been opposed by both Syria, where the river rises, and by Israel, which fears the effect of water loss.

The occupied lands, particularly the West Bank, but to a certain extent the Golan Heights, are increasingly vital to the water economy of Israel. The West Bank has become critical as a source of water for Israel and this consideration could well outweigh political and strategic factors. There are three major aquifers and, before 1967, Israel was exploiting two of these almost to the maximum by pumping from within its own borders. After 1967, Israeli control of the West Bank allowed access to the eastern aquifer with an estimated yield of 66mcm annually. The growth of Jewish settlements in the West Bank has of course increased the water requirement (Rowley, 1984), but there is a major disagreement about the actual amount available. While Israel considers the area to be self-sufficient, Jordan is convinced that there is a large surplus for use within Israel itself. Although the data cannot be checked, it has been asserted by some authorities that the increase in water consumption by Israel since 1967 has only been possible through territorial expansion (Naff and Matson, 1984).

Despite the completion of six reservoirs, Israel has gained comparatively little for its water budget directly from its occupation of the Golan Heights. Most water required in the area is taken from Lake Tiberias and thereby supplies for the remainder of the country are depleted. However, the Israeli presence in southern Lebanon has brought control of the sources of the upper Jordan. It has also given rise to Arab fears that Israel may try to divert the waters of the Litani into the Hasbani River through a tunnel, thereby providing Israel with an additional 500mcm of water annually.

The Jordan basin is well suited to integrated development, but all schemes proposed so far have failed as a result of the extreme enmity between the Arabs and Israel. The most significant was that proposed by Johnston, a United States special ambassador, to construct a comprehensive plan for the Jordan system. The starting point was the Main Plan which included a number of dams on the various tributaries, the reclamation of the Huleh marshes and gravity flow canals down both sides of the Jordan valley. Subsequent modifications were put forward, each apportioning different allocations to the four riparian states. Finally, the Johnston or Unified Plan received general acceptance, although ultimately it was not ratified by the Arab League Council. Nonetheless, the proposed allocation of water; Jordan 52%, Israel 36%, Syria 9% and Lebanon 3% has clearly been retained by all states as a guideline in the implementation of their own schemes. Thus, despite the failure to develop a multilateral approach, the Johnston Plan approached nearer total success than any other such proposed development in the Jordan basin.

Nevertheless, the various disputes within the region have not remained latent and have indeed led to more substantial geopolitical threats. An Arab summit meeting in 1964 resolved to divert the head waters of the Jordan tributaries outside Israel and Syria's subsequent diversion attempts resulted in a number of large scale border clashes. The outcome would have brought about the diversion of water from the Hasbani, Dan and Banias to the Yarmuk. The water conflict was effectively militarised in 1969 and Israeli raids on the East Ghor Canal resulted in severe damage.

Thus, the Jordan River has been the scene of more severe international conflicts over water than any other catchments in the Middle East and it remains by far the most likely flashpoint for the future. Not only has tension remained high but, more importantly, the water situation, particularly in Israel and Jordan, has continued to deteriorate. At present, it is estimated that Israel is consuming some 95% of all possible supplies. With the development of both states so crucially tied to water, since by far the major usage is in agriculture, it is only to be expected that further disputes will result.

5.3.4 Sub-surface Water

Competition for sub-surface water is more covert, but for both shallow and deep aquifers, the boundaries of which do not coincide with political frontiers, there is obvious potential for conflict **(see Fig.5.1)**. The example of Israel and the West Bank has already been discussed.

A similar case of sub-surface abstraction occurs in northwestern Oman. Ideal natural recharge conditions occur to the east of Buraimi and over time, these have produced particularly high yielding aquifers. During the past decade, over-pumping in Al Ayn (United Arab Emirates) has resulted in a dramatic decline in the water table beneath Buraimi. Within that period, a fall of at least 50 metres has been recorded. There are, of course, many similar geological configurations throughout the Middle East, providing an obvious potential for the recurrence of this problem. Geopolitical difficulties may even result from the large-scale use of water from deep aquifers. For example, the immense pumping programme in Libya must affect potential for development in the adjacent regions of Egypt (Odone, 1984).

5.4 Conclusion

Throughout most of the Middle East, with very low rainfall and therefore low recharge rates, there is a great shortage of naturally occurring fresh water. There are deep "fossil water" aquifers, but depletion of these is a contentious hydrological issue. In the case of both surface and ground water supplies, there are potential and actual geopolitical disputes, resulting from the non-coincidence of political and resource boundaries. Indeed, in the future it can be confidently predicted that water will become increasingly a key factor in confrontations over political frontiers.

References

Allan, J.A. (1985), 'Irrigated Agriculture in the Middle East: The Future' in Beaumont P. and McLachlan K. (eds), **Agricultural Development in the Middle East,** Chichester, Wiley.

Barney, G.O. (Study Director) (1980), **The Global 2000 Report to the President. The Technical Report, Volume Two,** The Council on Environmental quality and the Department of State, pp. 137-159.

Bodgener, J. (10 August, 1984), 'Oman Develops Skills Ancient and Modern' **Middle East Economic Digest,** p. 38.

Charnock, A. (10 August, 1984), 'Nile Schemes Bring Benefits and Problems' **Middle East Economic Digest,** pp 30 and 33.

Drysdale, A. & Blake, G.H. (1985), **The Middle East and North Africa: A Political Geography,** New York, Oxford University Press, p. 367.

Keen, M. (9 May, 1985), 'Cheaper, Purer Water from the Sun', **Water and Sewage,** pp. S14-S16.

Naff, T. & Matson, R.C. (1984), **Water in the Middle East. Conflict or Cooperation?,** Boulder, Westview Press, p. 236.

Odone, T. (10 August, 1984), 'Man-made River Brings Water to the People', **Middle East Economic Digest,** pp. 39-40.

Perara, J. (February 1981), 'Water Politics', **The Middle East,** pp. 47-54.

Rowley, G. (1984), **Israel into Palestine,** London, Mansell, p. 176.

Whittington, D. & Haynes, K.E. (1985), 'Nile Water for Whom?' in Beaumont P. and McLachlan K., (eds), **Agricultural Development in the Middle East,** Chichester, Wiley pp. 125-149.

Wilkinson, J.C. (1977), **Water and Tribal Settlement in South East Arabia, a Study of Aflaj of Oman,** OUP p. 276.

Fig. 1 MIDDLE EAST AND NORTH AFRICA: Water Resource Boundary Disputes

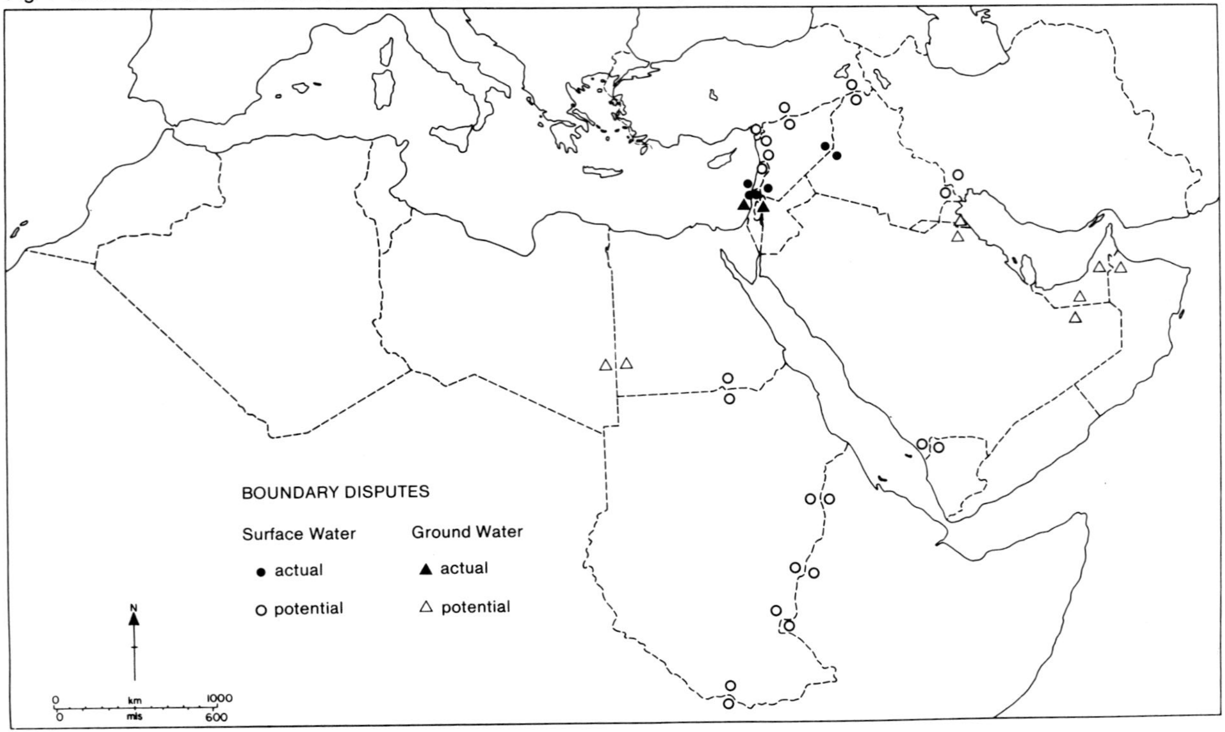

6. POSITIONAL DISPUTES AND THE TWENTIETH CENTURY UTILISATION OF THE SHATT AL-'ARAB - INTERNATIONAL RIGHT OR NEED ?

Richard N. Schofield

6.1 Introduction

The Shatt al-'Arab's dual role as a hotly-disputed international boundary and an economically-important, navigable river forms the focus of the following paper. After an initial section introduces this phenomenon, there follows subsequent, more detailed discussion of Iran's persistent request for Thalweg boundary delimitation in the estuary and a brief outline of the economic use and development of the river this century.

6.2 The Shatt al-'Arab River and Boundary

The Shatt al-'Arab is formed by the confluence of the Tigris and Euphrates at Qurna, from where it flows 118 miles (195 km) downstream into the Gulf. At around the half way point of its course, the Shatt al-'Arab's flow is considerably augmented by the Karun, Iran's longest and most important river. Indeed, two-thirds of the estuary's discharge at Fao, an outport on the Iraqi side of the river mouth and newsworthy recently because of its occupation by Iran in the Gulf War, derives from the Karun, (Cressey, 1958), **(see Fig. 6.1).** Arguments, particularly Iraqi, over their national rivers contribution to the Shatt al-'Arab's water discharge are viewed by some as evidence of the extent to which the river has become a political, rather than an economic or legal issue (Schofield 1986).

The international boundary between the modern and warring states of Iraq and Iran, and before them the Ottoman Empire and Persia has in whatever form, since 1847, been coincident with the lowest 65 miles (105 km) of the Shatt al-'Arab, **(see Fig. 6.2).** The development of both Abadan and Khorramshahr (formerly Muhammarah), Iran's principal ports on the river and until fairly recent times her most important nationally, has had to contend with periodic disquiet over the river boundary's status and alterations in alignment defined by Treaty. Basra, until recently Iraq's only dry-cargo port, while situated on the river is located some 20 miles or so north-west of the point where the international boundary leaves the Shatt al-'Arab to assume an on-land demarcation.

If the Iran-Iraq interface is something of a rarity in the Middle East in embracing a significant estuarine constituent, the Shatt al-'Arab remains exceptional in constituting the only navigable international river boundary in the region. Fishing and irrigation are undeniably important functions of the river. The evidence for the former lies in the fact that provision for fishing rights in the Shatt al-'Arab has been made in various agreements between Iran and Iraq. With respect to the latter, the tidal waters of the river irrigate one of the most important centres of date cultivation in the world with around 16 million trees growing on its banks. However, the increasingly saline nature of the estuary's water and more fundamentally the Gulf War have harmfully affected both phenomena.

6.3 Navigation in the Shatt al-'Arab

Navigation is, however, unquestionably the river's most important function. As early as the 1820s steamships were used on the Tigris, while the British East India Company had been able to establish agency facilities at Basra. Melamid (1968) noted the British government's concern at this time to find a faster route from Europe to India via Syria and the Tigris and Shatt al-'Arab river systems. The Naval Intelligence Division (1944) had referred to the river as 'broad' and 'stately' and throughout its course generally not difficult to navigate for large sea-going vessels, except at the bar near its mouth. Extensive investment in the dredging of bars affecting both the Shatt al-'Arab and the Karun had occurred in the late 1920s and subsequent improvements were speedily effected as a burgeoning oil industry necessitated increasing levels of river traffic and the accommodation of vessels of greater draught. Along with port developments and traffic levels within the estuary, the extent and rapidity with which bars have been breached are reliable indicators of the evolving economic value of the Shatt al-'Arab and its ports to both Iran and Iraq. These features are discussed in the paper's final section, with particular regard to the supposedly mitigating circumstances of the river's insecure boundary status.

As traditionally Iraq's only outlet to the sea, the importance of the navigable river need hardly be stressed. Today this has been agonisingly brought home to Iraq and with 80 vessels or more blocking the waterway she finds herself effectively landlocked - most disadvantageous even in peacetime.

6.4 Versailles Treaties of 1919

The premise for the dominant positional dispute over boundary alignment in the Shatt al-'Arab this century - that being the basic Iranian demand for a shift to Thalweg delimitation - is the Versailles Declaration of 1919. This specified that when fixing boundary alignment within a navigable river the Thalweg principle be adopted as the general rule. Arbitrarily one might define such a principle by stating that the boundary must run through the mid-channel of the river. However the Thalweg is a vague and ambiguous term which possesses no one definition in international law, a fact that has aggravated the prevalence of many positional disputes. If rivers were not navigable, the Versailles Treaties prescribed that the boundary run down the median line (**medium filum aquae**) following all turnings of course and remaining equidistant from both banks.

The problem for the Shatt al-'Arab lay in the fact that by the time of Versailles (1919) her boundary alignment had already evolved through two treaties in international law, (Erzerum, 1847 and the Constantinople Protocol of 1913), **(see Figs. 6.2 & 6.3).** This was less due to the mutual requirements of the Ottoman Empire and Persia than the Strategic and economic concerns of the Great Powers, Great Britain and Russia. In the same way that Boggs (1940) described a 'complex' land boundary - i.e. that combining physical, man-made features and arbitrary geometric lines - one was increasingly able to describe a 'complex' river boundary for the Shatt al-'Arab after 1913 and 1937 especially (Bouchez 1963). The Tehran Treaty of that year did not heed the prescriptions of the Versailles Declaration but rather the positional changes it introduced left the navigable river subject to median line and Thalweg delimitation at various stages in its course. However most of the river was still under Iraqi sovereignty, **(see Fig. 6.4).**

6.5 Boundary Evolution in the Shatt al-'Arab

In the space of 130 years the Shatt al-Arab's status changed from that of total Ottoman sovereignty before 1847 (Ismael 1982) to the 'international norm' for navigable rivers of Thalweg delimitation in the 1975 Algiers Accords, **(see Fig. 6.5).** It was anything but a smooth, continuous transition however and the stipulations of any treaty regarding the estuarine frontier always proved disputed and impermanent. Increasingly, during the course of this century, there is a case for suggesting that the Shatt al-'Arab boundary has been used as the physical expression of wider state rivalries between Iran and Iraq. This reminds us of Ancel's (1938; 196) classic comment:

> "There are no problems of boundaries. There are only problems of nations".

For a discussion of the powers directly or indirectly responsible for each treaty defining territorial change in the Shatt al-'Arab and the predominant interests governing the creation and content of these formal agreements see Melamid (1968), Sevian (1968) and Schofield (1986).

Whilst a long-standing Iranian 'positional' dispute is being considered first and foremost here, there is sometimes a danger in labelling one particular interface as subject to one particular boundary dispute classification. Prescott (1978) classifies the history of dispute over the Shatt al-'Arab boundary as 'functional'. This is surprising when one considers that he has always seemed at pains to point out that the resolution of such a category of dispute does not require an actual change in boundary alignment. This is not to deny that functional arguments between the riparian states have taken place. In Iran's case for example, accusations over the mishandling of dues by the Iraqi-controlled Basra Port Authority have usually followed impasses in their attempts to secure the prime positional objective - Thalweg delimitation in the estuary. Iraq, in 1975, felt compelled to concede to Thalweg delimitation because of desperate overriding circumstances. Once internal conditions appeared to have ameliorated however, Iraq would be forced to enact a 'territorial' dispute if it wished to regain sovereignty over the Shatt al-'Arab. This she did in September 1980, as a prelude to the Gulf War, on the premise that historically the river had always been an Iraqi national river. This all suggests that boundary disputes are not susceptible to just one classification and that it can prove far more illuminating to highlight the most important, effectual and prevalent entered over its lifespan.

6.6 History of Iranian Thalweg Claim

It is largely the contention of this paper that the economic use and development of the Shatt al-'Arab has proceeded almost independently of whatever the estuary's precise boundary alignment has been at any one time. Nevertheless, one would expect that for a navigable boundary river, the Thalweg would prove the most practical and functional form of delimitation. The functional validity of the principle is open to question however. Where a choice has had to be made between the centre of a stream (**median line**) and the centre of the navigable channel (**Thalweg**), decisive emphasis has generally been placed upon the fact that if the former were defined as the boundary line, the whole track of navigation would be included in the territory of one state to the exclusion of the other. Yet, with the latter, if vessels were to keep to the right (as is customary), the bulk of upstream navigation is likely to pass entirely within the waters of the right-hand riparian state. Vessels proceeding downstream would find themselves within the jurisdiction of both states if sailing directly over the Thalweg, or just within or without the jurisdiction of one to the exclusion of the other (Lauterpacht 1960). A few commentators have suggested that the most suitable delimitation for a divided international river would be joint sovereignty. Each riparian's sovereign territorial limit would be defined by the river bank at the mean low water mark (Lauterpacht 1960; Schofield 1986).

According to unconfirmed sources the first boundary arrangement in the Shatt al-'Arab to approximate to the Thalweg principle occurred in the mid-nineteenth century when:

> "Inhabitants of the southern border region were understood to have reached an informal agreement under which the boundary was regarded as running down the centre of the Shatt al-'Arab, with both sides controlling navigation". (Day 1982; 216)

Perhaps forgivably, preceding the Versailles Declaration, the terms **median line** and **Thalweg** were mistakenly regarded as interchangeable in the 1913 Protocol of Constantinople and the Proces Verbaux of the following year. It was more surprising that they continued to be confused in the text of the 1937 Tehran Treaty, especially as this agreement had been reached some years after the first formal request by Iran for Iraq to accept Thalweg delimitation in the Shatt al-'Arab (Schofield 1986).

Six months prior to Iraq's formal admission to the League of Nations as an independent state on 10 March 1932, King Faisal I, while visiting Tehran, was confronted with an Iranian request that the boundary in the river be adjusted to the 'international norm' of Thalweg delimitation. This set in motion a destabilisation in relations between the two states and both Ismael (1982) and Ramazani (1966) attest to the fact that the status of the river was the basic issue to dominate Iran-Iraq relations in the early 1930s. Accusations and counter-accusations of border violations by both sides and blatant Iranian disregard of the pilots and navigation signals of the Basra Port Authority were examples of the numerous functional disputes which had increasingly occurred since Reza Shah's forceful occupation of Khuzestan in the mid-1920s. Tensions reached such a pitch that by 1934 Iraq felt compelled to bring the attention of the League of Nations to the growing disquiet in the estuary.

6.6.1 The 1934 Presentations to the League of Nations

Only a matter of days had passed before Iran responded by stating that she, too, was willing to present her views to the Council. The diplomatic instruments that Iraq could use to support its case, particularly the 1847 Treaty of Erzerum and the 1913 Constantinople Protocol, were the very instruments which Iran provided to illustrate the injustice of the boundary's evolution and her own manipulation by the Great Imperial powers of Great Britain and Russia. The clause in the Erzerum Treaty which specified that:

> " ... Persian vessels shall have the right to navigate freely without let or hindrance on the Shatt al-'Arab ...". (League of Nations Official Journal Feb. 1935)

found a new interpretation before the Council. The Iranians were to suggest it indicated that the riparian states will have an equal right of sovereignty as far as the middle of the river (U.S. Intelligence Report 1960). In a counterplea to maintain the **status quo**, Iraq introduced an argument of Equity. This stated that, while Iran possessed a coastline of 1,250 miles with many ports and anchorages, the Shatt al-'Arab constituted Iraq's only access to the sea, whose maintenance as such had to be safeguarded to ensure the functioning of Basra. Each side's stubbornness meant that the respective cases were dropped and an Iraqi proposal that the matter be referred to the Permanent Court of International Justice was vetoed by Iran.

6.6.2 The 1937 Tehran Treaty

It was some surprise, therefore, that the states thereafter resumed direct negotiations which were to culminate in their signature in Tehran of the Frontier Treaty of 1937. This was to introduce a Thalweg delimitation but only for the six miles alongside Abadan anchorage, supposedly to provide the fast-developing oil port with the sovereign waters in which increasing numbers of vessels could be securely berthed and laden. Rights of free navigation for Persian vessels were redefined but effectively re-affirmed in the second paragraph of Article Four (Lauterpacht 1963). The legal ambiguities of 'free navigation' were to become a problem once again in the late 1950s when Iran wanted to considerably expand its port of Khosrowabad.

Various observers (not always totally impartial) viewed the results of the Treaty in different ways. Ramazani (1966) regarded both sides to have made significant concessions while Ismael (1982) suggested that the newly-appointed Iraq government under Bakr Sidky had succumbed to Iranian pressure. Iran voiced sporadic discontent in the years immediately following the Treaty as she was increasingly to regard the instrument as a device for maintaining British strategic interests (Niblock 1982). This argument may well hold more water than one might initially suppose to be the case.

In 1936 a possible Thalweg delimitation in the Shatt al-'Arab had become the subject of impassioned discussions in London between various Government Departments. The Foreign Office regarded the adoption of such a principle as the only feasible long-term resolution. The Admiralty and Colonel Ward, British Administrator of the Port of Basra opposed such a development and their argument was to eventually win the day. What had united the three wrangling parties however was their determination to maintain Great Britain's strategic and economic interests in the region. There had been evidence that at one time in the mid-1930s Iraq was considering the acceptance of a Thalweg delimitation to the whole Shatt al-'Arab. The fear that a compromise struck between the riparians might neglect British interests was paramount in London's thinking. As the despatch from the Foreign Office to the Admiralty of December 10, 1936, illustrates:

> "We may, in fact, easily find ourselves in a position where the Iraqis are prepared to meet the Persian desire for a Thalweg frontier, and where both parties will be able legitimately to represent that we alone are standing in the way of a settlement". (F.O. 371/20039/235 - Quoted in Sanghri, 1969)

Britain's **de facto** control of the river area and neighbouring Iranian oilfields to the east certainly continued until the end of the Second World War and only really disappeared with the oil crisis of the early 1950s and the accession of Qasim to power in the Iraqi coup d'etat of July 14, 1958. Perhaps the reason why the economic development of the estuary had been so smooth and continuous up to this juncture was because Britain, with its appreciable level of control, would not allow what was generally the rhetoric of riparian claims and counterclaims to develop into anything more serious than minor functional skirmishes. Under Britain's last phase of guardianship, traffic on the Shatt al-'Arab steadily increased while remaining under the sole control of her appointed Basra authorities. The connections of the port of Khorramshahr to the Iranian railroad system had greatly aided these developments.

6.6.3 Hardening of Stances - Late 1950s until 1975

Despite Mossadegh's nationalist aspirations the boundary dispute over the Shatt al-'Arab was to remain dormant between 1951 and 1954, though the oil nationalisation procedures naturally led to a substantial decrease in river traffic. Once the young Muhammad Reza Pahlavi had been reinstated Iran started to stress the functional economic argument regarding the collection and utilisation of taxes by the Basra authorities rather than purely reiterating their basic positional claim for a shift to Thalweg delimitation (Melamid 1968).

The Iraqi Revolution of 1958 marked the beginning of a period in which disputes over the Shatt al-'Arab regime or its status had increasingly to be viewed in the context of wider and fast-developing state rivalries such as religion, ethnicity and ideologies. External powers had generally lost their political leverage in the area, though their economic concerns continued to develop massively through oil interests. In the muscle-flexing of this period, Qasim renounced the 1937 Treaty and claimed Iraqi sovereignty over the whole of the Shatt al-'Arab. Iran once again clamoured for Thalweg delimitation. With the backcloth of stability provided by a strong imperial presence and influence now gone, the question of the river would be increasingly subject to downturns in relations between the politically immature states of Iran and Iraq and any changes in the balance of power between them.

Chubin and Zabih (1974) characterise the period between the late 1950s and the mid-1970s as one in which Iran-Iraq relations were almost "uniformly poor". They also recognise a marked shift in power and initiative from Iraq to Iran as having occurred, especially when considering the Shatt al-'Arab waterway, over which Iran took a hardening stand from 1969. This is illustrated perfectly by incidents in 1961 and 1969, during which each power tested the other out.

Having met with Iraqi inflexibility concerning a demand for sole, uninterrupted control of port operations at Abadan as well as other ports, Iran, on February 10, 1961, issued a ruling under which vessels entering Iranian waters or ports would be guarded by Iranian pilots. Iraq retorted by effecting a Basra Port Authority strike, which resulted in a nine-week virtual standstill in the waterway. Iraq proved unwilling to compromise and Iran unwilling to violate Iraq's territorial waters and the latter soon agreed to return to a **status quo ante.**

On April 15, 1969, following Iraqi claims to Khuzestan and reassertions of the Iranian stance on Thalweg delimitation, Iraq announced that all ships flying the Iranian flag would be required to lower them in the estuary and that all Iranian nationals on board ships in Iraqi water should disembark. The Iranians responded by abrogating the 1937 Treaty with the Shah denouncing Iraq's failure to live by its provisions. The now substantial Iranian armed forces were alerted and all Iranian and Iran-bound shipping was to receive a military escort. Facing such overwhelming odds, Iraq backed down.

6.6.4 The Algiers Accords of 1975

The 1975 Agreement was the first genuinely bilateral boundary delimitation Treaty struck between the two neighbouring powers. This instrument was the latest in a succession of many, yet the first to specify the 'international norm' of Thalweg delimitation for a navigable river. It might have been supposed that the boundary had evolved in international law to finally satisfy the mutual requirements of its riparian states. However, its formulation and signature were more the result of one destabilised and vulnerable state pitted against its momentarily stronger counterpart.

It was in the early 1970s with the Shah's strong support for the guerilla operations of the Iraqi Kurdish leader Mulla Mustafa Barzani, that the recognised danger of the political disintegration of an ethnically heterogeneous state was approaching reality. The concession was above all a peace signal from a desperate and weak Iraq, for which a conclusion to the Kurdish problem temporarily had to override all other considerations.

Saddam Hussein unilaterally abrogated the Algiers Agreement of 1975 on September 17, 1980. This supposedly 'final' Thalweg delimitation had once again been rendered short-lived by perceptions of the balance of power between Iran and Iraq as having altered, even if these perceptions have subsequently proved wildly over-optimistic.

6.7 Economic Use and Development of the Shatt Al-'Arab in the Twentieth Century

Three phases relating to the scale of port operations and their location can be identified. The first relates to the situation pertaining before and during World War I, when Basra's initial port of Ashar, while economically important locally, had only been internationally of note in its military function. A second phase saw the rapid development of Basra, Khorramshahr and Abadan, the latter particularly due to the increasing export levels of Iranian crude oil and its products. A third phase, initially due to the technical shortcomings of the upstream ports, has seen a stabilisation or even decrease in the level of river traffic, essentially in the last one and a half decades, and a migration of port functions to the Head of the Gulf out of the disputed estuary proper. This tendency has been reinforced by the closure of the Shatt al-'Arab during the Gulf War.

6.7.1 Pre-1945

Basra, known as the 'Venice of the East' because of its situation as a focal point for a substantial canal and water channel network, had become an important trading port in the time of the Abbasid Caliphate (9th - 10th Century). However, on the eve of the First World War she was by no means an advanced port.

Ashar consisted of only three customs sheds, possessed handling facilities for only two non-native vessels at any one time and all loading and unloading was by lighterage in mid-stream (Byford 1937).

There was to be a locational change in the nucleus of Basra Port when the city became the base for a large British Expeditionary Force in Mesopotamia during the Great War (Godwin and Mitchell 1982). This new development at Maqil, 6 miles upstream of Basra, gathered pace after the Basra Port Declaration of 1919, issued by the resident British Military authorities there.

The 1920s, as was stated in the introduction to this paper, saw extensive investment in the dredging and deepening of the Shatt al-'Arab's main navigation channel, in order that the growth of its river ports would not, in future, be constrained.

That such endeavours were successful is evidenced by the dramatic expansion of maritime trade from pre-war days to the late 1930s. The Basra Port under British **de facto** control was now very much a region rather than a single port embracing, as well as Maqil, the Iranian ports of Khorramshahr and Abadan and the Iraqi Fao depot, later to become Basra's outport, (Byford 1937), **(see Fig. 6.6).** Khorramshahr had already become the principal commercial port of Iran while Abadan was responsible for the vast majority of tonnage entering and leaving the waterway. She possessed 10 jetties and by 1937 the export of crude oil and its refined products from Abadan was approaching the order of 7 million ton per annum. Fao, at this stage, consisted only of a branch office for the collection of dues on sailing craft which did not proceed so far upstream as Basra (Schofield 1986).

Whereas the total tonnage of ships entering Basra in pre-war days had been estimated at 400,000 ton (Byford 1937), the total entering the aforementioned Basra Port region had increased to around 10,000,000 ton by the fiscal year of 1935/1936. Abadan's export of oil and oil products dominated this figure. Obviously the positional and functional disputes over the river at this point, aired most notably before the Council of the League of Nations, had had little harmful effect on the Iranian oil port's rapid development. There was a discernible and constant increase in tonnage exported from Abadan throughout this decade, both before and after the Tehran 1937 Treaty.

This casts doubt on Melamid's (1968) contention that the prevailing atmosphere of mistrust and instability between the two states over the river had been responsible in the 1930s for the location of the Trans-Iranian railroad terminus at Bandar Shahpur (now Khomeini) at the head of Khor Musa, rather than at Khorramshahr on the Shatt al-'Arab. The high cost of extending the railway from Ahwaz to Khorramshahr could no doubt have been a factor, but this was precisely what the British were to do during World War Two.

6.7.2 1945-1975

The early 1950s were notable for the inauguration of oil production in Iraq's southern fields and its subsequent export from ports in the Shatt al-'Arab region - most notably Fao. While Maqil and Khorramshahr were experiencing steady growth, oil was first exported from Fao in December 1951. She was directly linked to Zubair and Rumaila oilfields by pipeline. Fao's development was so spectacular that by the end of the decade her oil exports were approaching those of Abadan, **(see Table 6.1).** Such trends would also appear to refute the oft-voiced Iranian claim of the time that over seventy-five per cent of ships entering the Shatt al-'Arab were bound for Iranian ports.

The late 1950s had also seen the construction of Iraq's first deep sea offshore Gulf terminal - Khor al Amaya. South of Fao and out of the estuary proper, Melamid (1968) contended that its location had been chosen because of an impasse in attempts to resolve the boundary question in the Shatt al-'Arab. Yet there is a compelling case for suggesting that the reasons for its construction and that of its later and larger counterpart, Mina al-Bakr, are the same as those which were to dictate Abadan's demise and Kharg Island's rise to eminence. For increasingly from 1960 onwards the continued increase in tanker size (up to 300,000 dead weight ton - d.w.t.) rendered the Iranian terminals at Abadan largely obsolete.

After the Iranian oil crisis of the early 1950s the international consortium of oil companies began to construct new deep sea terminals on the Iranian Kharg Island in the Gulf. As well as the need to accommodate larger vessels, this relocation of export shipments was also dictated by the inauguration of oilfields a considerable distance from Abadan. The decline in traffic on the Khor Musa led to the conversion of Bandar Mashur into a terminal for the loading of petroleum products piped from Abadan in

TABLE 6.1 : TOTAL IMPORT & EXPORT FIGURES FOR THE IRANIAN AND IRAQI PORTS ON THE SHATT AL-'ARAB

Name of Port	Year	In 000s metric tons Imports	Exports	Total
Iraqi Ports				
Basra (Maqil)	1957	668	356	1,024
Basra (Maqil)	1958	651	473	1,124
Fao	1956/57 1	-	7,968 2	7,968
Fao	1957/58 1	-	9,487 2	9,487
Iranian Ports				
Khorramshahr	1956/57 1	374	339	713
Khorramshahr	1957/58 1	435	216	651
Abadan	1956/57 1	251	10,133	10,348
Abadan	1957/58 1	175	11,310	11,485

Notes : 1 - Relates to fiscal year begining April 1
2 - Crude oil only

Source : After, The Dispute over the Shatt al-'Arab, p.32, Intelligence Report No.8208, Bureau of Intelligence and Researcxh, Washington D.C. (22.1.1960)

an effort to rejuvenate activity in the water outlet. These factors effectively removed the last Iranian petroleum activity from the Shatt al-'Arab. Immediately preceding the Gulf War, Abadan's oil-loading facilities were used only for coastal tankers and the bunkering of freighters bound for Khorramshahr or Basra (Riethmuller 1979).

In his incisive paper of 1968, Melamid also suggested that with the construction of new Iranian dry cargo ports on the Gulf such as Bandar Abbas and the expansion of those already existing such as Bandar Shahpur, the volume of traffic handled in Khorramshahr would also be substantially reduced in future years. Unfortunately Iraq had no alternative coastline on which like developments might be accommodated. She had to keep faith with and gradually improve Basra. Considerable investment in new berthage for each of Basra's ports of Maqil and Ashar resulted in exports increasing from an annual 400,000 ton to 1.4 million ton over the period 1968-1977. By the later 1970s it was becoming obvious however that, even with these improvements, Basra's capacity had reached saturation point. Further growth in maritime trade would have to be accommodated by Umm Qasr, Iraq's second dry cargo port near the mouth of the Khor Zubair. Though opened in 1967, Umm Qasr was only really to become a feasible option in 1976, when she was linked directly to Basra by good road and rail connections. **Table 6.2** illustrates the extent to which Iraq had intended to move maritime trade out of the Shatt al-'Arab through the intensive development of Umm Qasr, but the Gulf War was to cut short any such ambitions.

Melamid's (1968) prediction over Khorramshahr's future was to be proved correct in the 1970s. As **Table 6.3** shows, of all the Gulf's dry-cargo ports, only Khorramshahr experienced significant decline. Inspection of **Table 6.2**, detailing existing berths and those planned or under construction, further highlights such a point and also reinforces the evidence that migration of port functions from the Shatt al-'Arab to the Gulf was not to be confined to Iran, but also extended to Iraq.

The substantial decline in the importance of Khorramshahr, as opposed to the increasing buoyancy of Bandar Khomeini (Shahpur) and Bandar Abbas, is ironic when one realises that it was only in 1975 that Iran's age-old demand for Thalweg boundary alignment was met. Though this delimitation was only to last five years - **de facto** and **de jure** - Khorramshahr's decline clearly preceded and was independent of the cause of the end to this short lived boundary line - the heightening tensions between the Shatt al-'Arab's two riparians which erupted in full-scale war in 1980.

To conclude, it is pertinent to review Melamid's (1968; 357) comment:

TABLE 6.2
DEEP WATER BERTHAGE IN IRAQI & IRANIAN COMMERCIAL PORTS

Port	Existing Number	Planned or under Construction	Total
Iraq			
Basra	20	0	20
Umm Qasr	4	15	19
Iran			
Khorramshahr	13.5	0	13.5
Bandar Shahpur (Khomeini)	12	27	39
Bandar Abbas	8	24	32

Source : After Grey Mackenzie Ltd, The Gulf Pattern 1977-82, p.94, (London), 1976

TABLE 6.3 : IRAQI AND IRANIAN COMMERICAL PORTS
DRY CARGO IMPORTS AND EXPORTS (in 000s ton deadweight)

	Imports			Exports		
	1975	1976	1977	1975	1976	1977
Iraq						
Basra	3,029	3,365	1	1,161	1,090	1
Umm Qasr	208	65	1	266	189	1
Iran						
Khorramshahr	3,638	4,694	3,571	268	197	187
Bandar Shahpur (Khomeini)	3,563	4,194	5,451	249	204	278
Bandar Abbas	1,796	2,508	3,012	174	265	187

Note : 1 - Figures not available

Source : After Grey Mackenzie Ltd, The Gulf Pattern 1977-82, p.93, (London), 1976

"The relocation of petroleum export and dry cargo activities from the Shatt al-'Arab appears to have reduced significantly the intensity of the river boundary dispute".

Certainly one might state that with the changing geography of operations, the economic importance of the river **per se** decreased for both riparians. Yet periodic dissatisfaction with the various river boundary delimitations utilised up until the 1975 adoption of the Thalweg had not demonstrably affected the growth of dry-cargo and oil-loading ports on either side of the estuary in an adverse manner. It was as if very strong market forces had to a large extent overriden all other factors such as the well-aired positional and functional disputes entered over the river in the last 50 to 60 years. Each delimitation used on the river this century has not, barring the occasional flashpoint, practically affected the river's economic utilisation and development.

6.4 References

Ancel, J., 1936, **Les Frontieres,** Paris.

Boggs, S.W., 1940, **International Boundaries: a Study of Boundary Functions and Problems,** New York.

Bouchez, L.J., 1963, 'The Fixing of Boundaries in International Boundary Rivers', in **International and Comparative Law Quarterly,** July 1963, pp. 789-817.

Byford, C., 1937, **The Port of Basrah, Basrah, Iraq,** London, Published under the Authority of the Port of Basrah Directorate.

Chubin, S. and Zabih, S., 1974, **The Foreign Relations of Iran: A Developing State in a Zone of Great Power Conflict,** Berkeley, University of California Press.

Cressey, G.B., 1958, 'The Shatt al-'Arab Basin', in **Middle East Journal,** Geographical Review, Autumn 1958, pp. 448-460.

Day, A.J. (ed.), 1982, **Border and Territorial Disputes,** A Keesing's Reference Publication, London, Longmans.

Godwin, C. and Mitchell, J.K.M., 1982, **Ports of the Arabian Peninsula** (unpublished), University of Durham.

Ismael, T.Y., 1982, **Iraq and Iran: Roots of Conflict,** Syracuse,

Syracuse University Press.

Lauterpacht, E., 1960, 'River Boundaries: Legal Aspects of the Shatt al-'Arab Frontier', in **International and Comparative Law Quarterly**, Vol. 9, 4th Series, pp. 208-236.

League of Nations, 1935, **Official Journal**, Feb. 1935, pp. 197-226.

Melamid, A., 1968, 'The Shatt al-'Arab Boundary Dispute', in **Middle East Journal**, Washington D.C., Vol. 22, pp. 351-357.

Naval Intelligence Division, 1944, **Iraq and the Persian Gulf**, London, Geographical Handbook Series, BR 524.

Niblock, T. (ed.), 1982, **Iraq: the Contemporary State**, London, Croom Helm.

Prescott, J.R.V., 1978, **Boundaries and Frontiers**, London, Croom Helm.

Ramazani, R.K., 1966, **The Foreign Policy of Iran: A Developing Nation in World Affairs, 1500-1941**, Charlottesville, University Press of Virginia.

Riethmuller, J. (ed.) 1979, **Ports of the World**, 32nd Edition, London, Benn.

Sanghvi, R., 1969, **Shatt al-'Arab: The Facts Behind the Issue**, London, Transorient.

Schofield, R.N., 1986, **Evolution of the Shatt al-'Arab Boundary Dispute**, Wisbech, Menas Press.

Sevian, V.J., 1968, 'Evolution of the Boundary between Iraq and Iran,, in **Essays in Political Geography**, London, Methuen, pp. 211-223.

United States Bureau of Intelligence and Research, 1960, **The Dispute over the Shatt al-'Arab**, Intelligence Report No. 8208, Washington D.C., January 22nd, 1960.

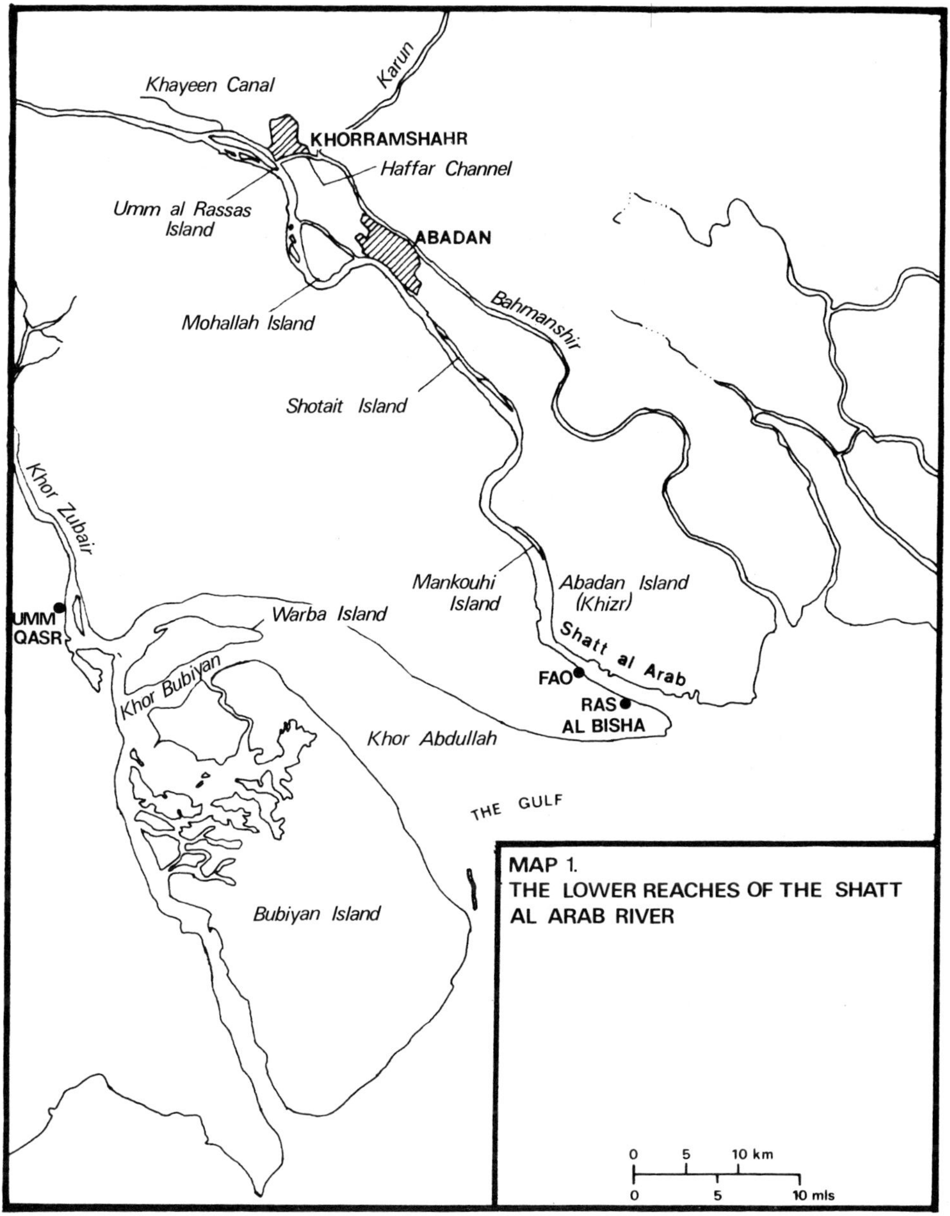

MAP 1.
THE LOWER REACHES OF THE SHATT AL ARAB RIVER

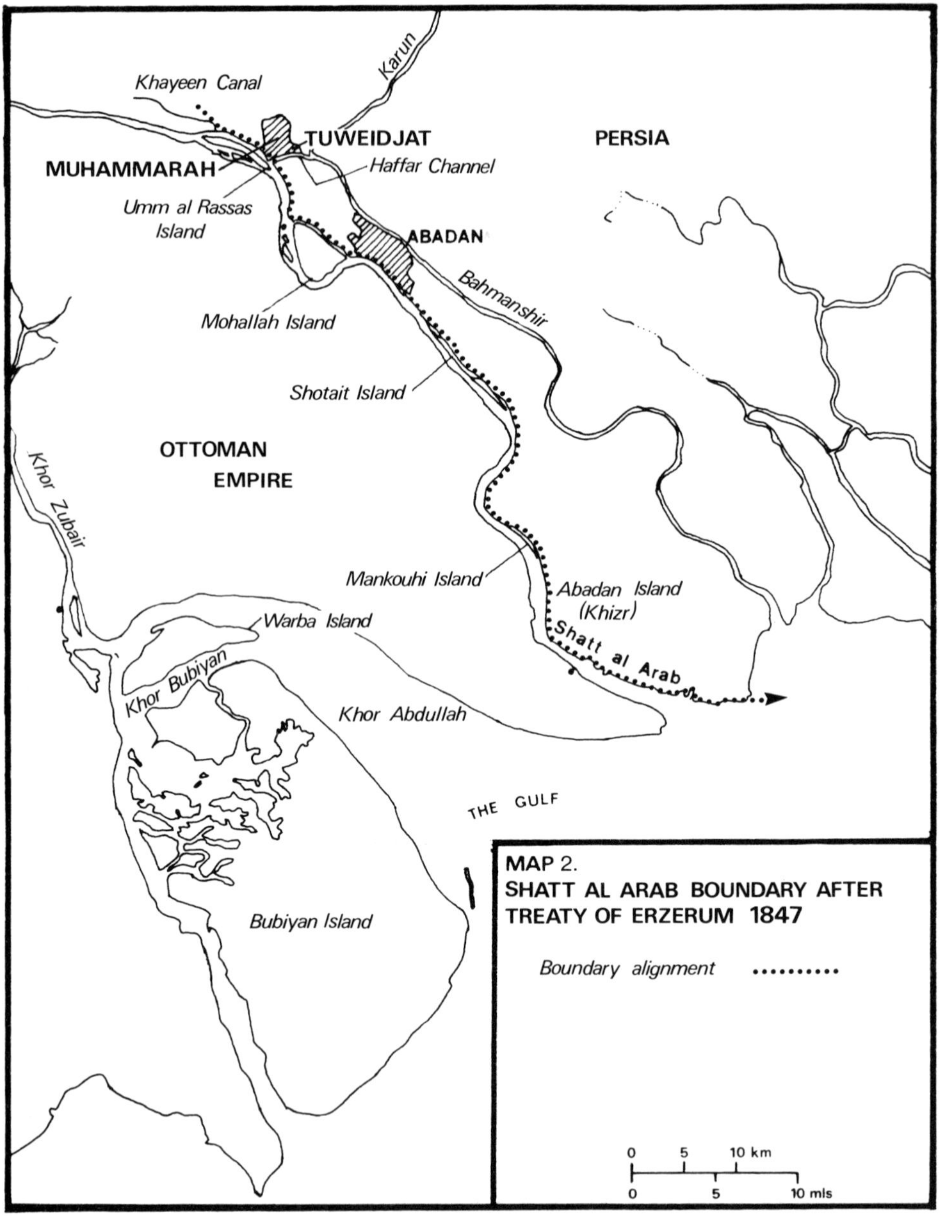
Karun
Khayeen Canal
TUWEIDJAT
PERSIA
MUHAMMARAH
Haffar Channel
Umm al Rassas
Island
ABADAN
Bahmanshir
Mohallah Island
Shotait Island
OTTOMAN
EMPIRE
Khor Zubair
Mankouhi Island
Abadan Island
(Khizr)
Shatt al Arab
Warba Island
Khor Bubiyan
Khor Abdullah
THE GULF
Bubiyan Island
MAP 2.
SHATT AL ARAB BOUNDARY AFTER
TREATY OF ERZERUM 1847
Boundary alignment
0 5 10 km
0 5 10 mls

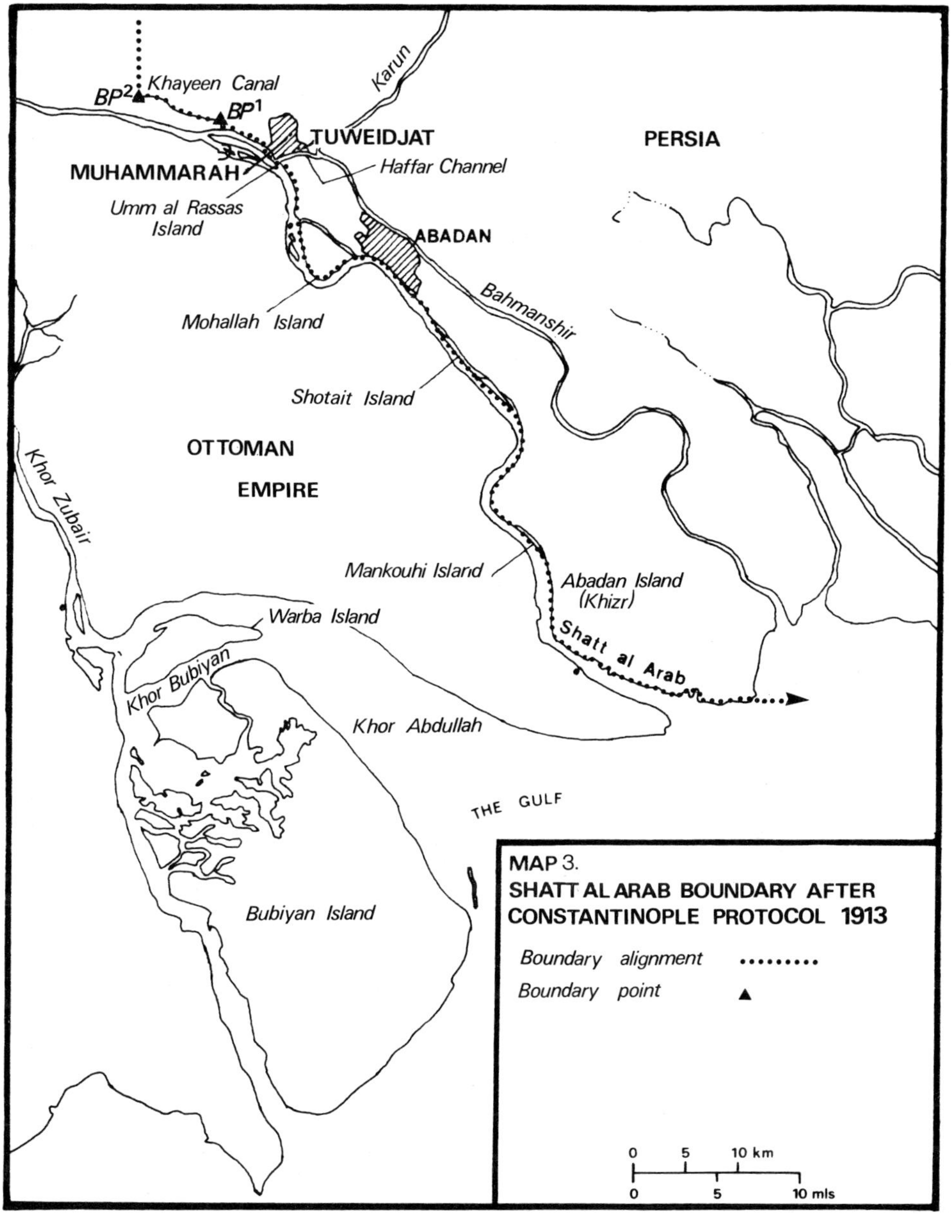
BP2
Khayeen Canal
BP1
Karun
TUWEIDJAT
PERSIA
MUHAMMARAH
Haffar Channel
Umm al Rassas Island
ABADAN
Bahmanshir
Mohallah Island
Shotait Island
OTTOMAN EMPIRE
Khor Zubair
Mankouhi Island
Abadan Island (Khizr)
Warba Island
Shatt al Arab
Khor Bubiyan
Khor Abdullah
THE GULF
Bubiyan Island
MAP 3.
SHATT AL ARAB BOUNDARY AFTER CONSTANTINOPLE PROTOCOL 1913
Boundary alignment
Boundary point
0 5 10 km
0 5 10 mls

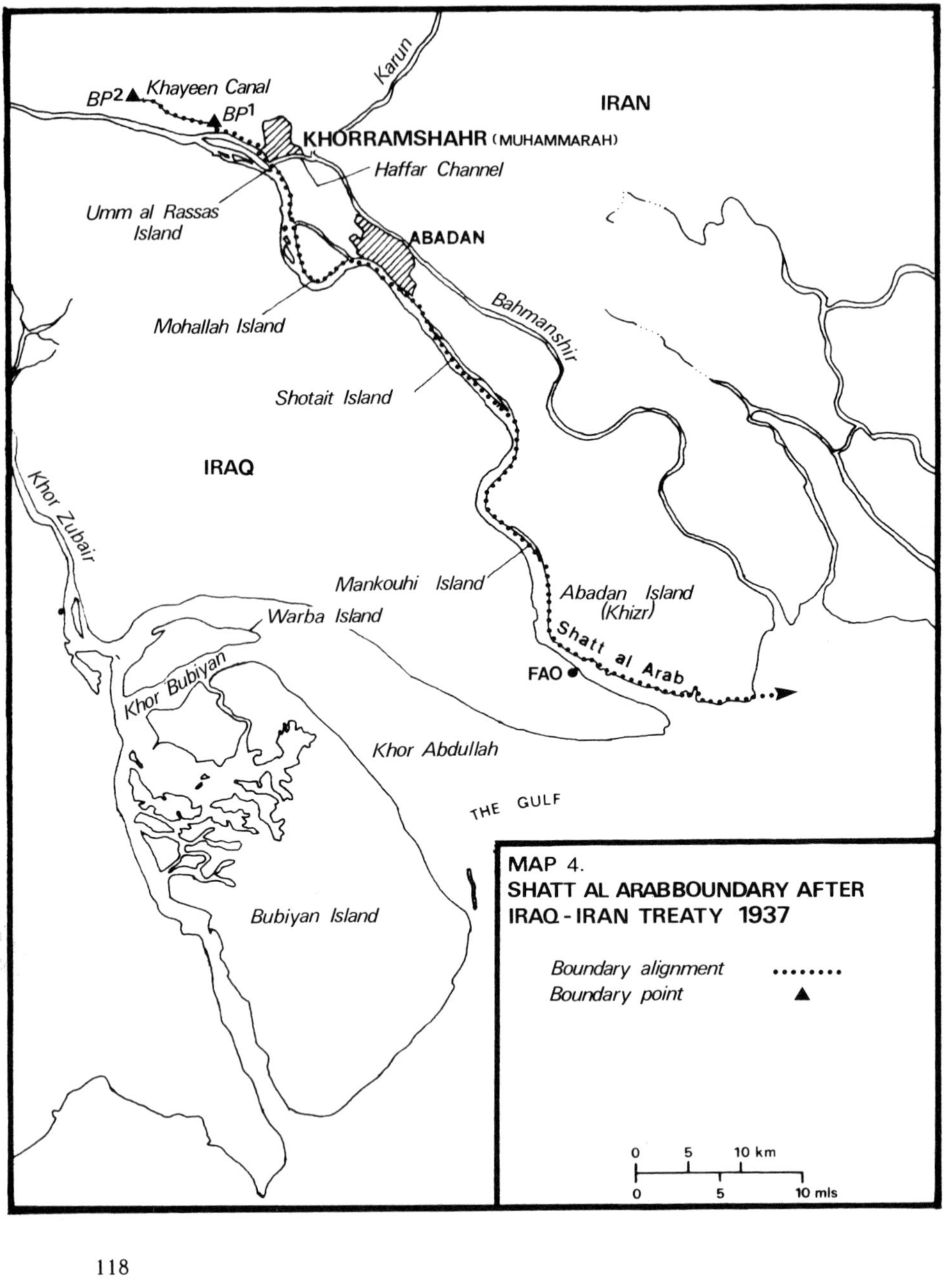
BP2
Khayeen Canal
BP1
Karun
IRAN
KHORRAMSHAHR (MUHAMMARAH)
Haffar Channel
Umm al Rassas Island
ABADAN
Mohallah Island
Bahmanshir
Shotait Island
IRAQ
Khor Zubair
Mankouhi Island
Abadan Island (Khizr)
Warba Island
Shatt al Arab
FAO
Khor Bubiyan
Khor Abdullah
THE GULF
Bubiyan Island
MAP 4.
SHATT AL ARABBOUNDARY AFTER
IRAQ-IRAN TREATY 1937
Boundary alignment
Boundary point
0 5 10 km
0 5 10 mls

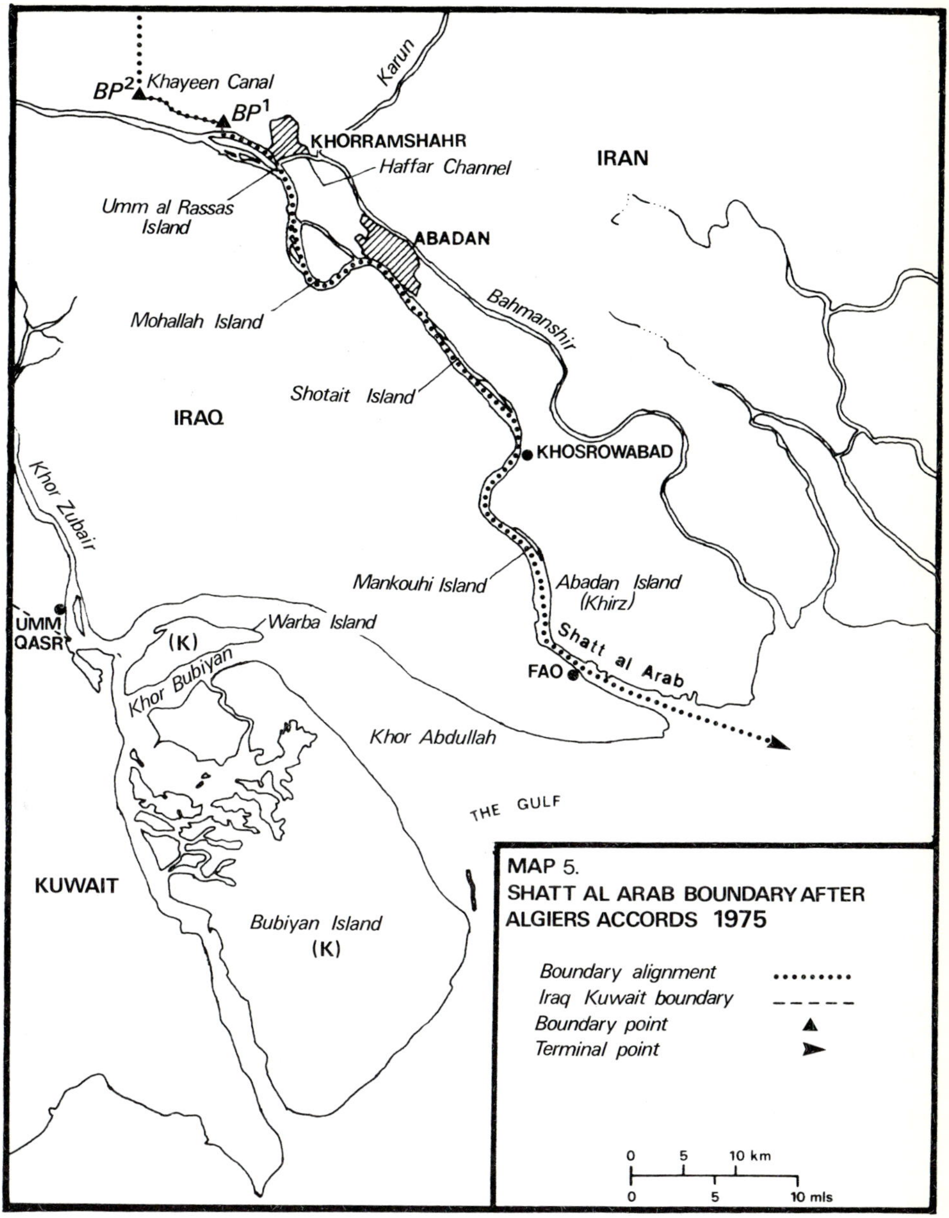

BP2
Khayeen Canal
BP1
Karun
KHORRAMSHAHR
Haffar Channel
IRAN
Umm al Rassas Island
ABADAN
Mohallah Island
Bahmanshir
Shotait Island
IRAQ
KHOSROWABAD
Khor Zubair
Mankouhi Island
Abadan Island (Khirz)
UMM QASR
(K)
Warba Island
Shatt al Arab
Khor Bubiyan
FAO
Khor Abdullah
THE GULF
KUWAIT
Bubiyan Island (K)
MAP 5.
SHATT AL ARAB BOUNDARY AFTER ALGIERS ACCORDS 1975
Boundary alignment
Iraq Kuwait boundary
Boundary point
Terminal point
0 5 10 km
0 5 10 mls

Map 6

PORTS IN THE SHATT AL ARAB REGION

(excluding Bandar Khomeini & Khor Musa)

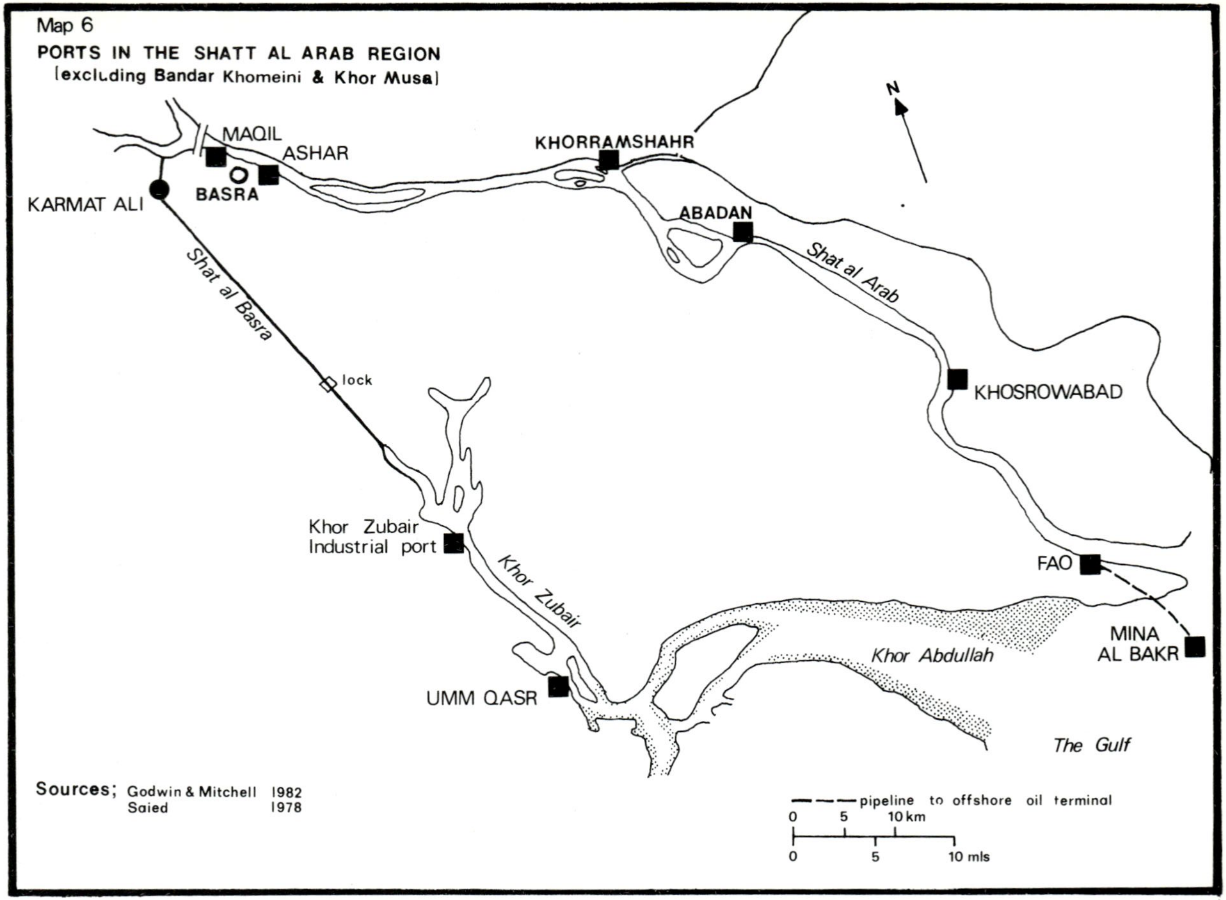

Sources; Godwin & Mitchell 1982
Saied 1978

7. MARITIME BOUNDARIES OF THE MIDDLE EAST AND NORTH AFRICA

Gerald H. Blake

7.1 Introduction

The partitioning of the oceans began in earnest little more than two decades ago, largely as a result of the quest for offshore hydrocarbon deposits. The process of maritime boundary delimitation is far from complete, and it may be several decades before the world maritime political map is finalised. There are approximately 380 potential maritime boundaries worldwide, about 100 of which have been agreed so far (Smith 1982; 1-22). The highest concentration of potential boundaries is in the Caribbean region, followed by the oceans and seas which fringe the Middle East and North Africa where at least 62 potential boundaries have been identified **(see Table 7.1).** It should be noted that **Table 7.1** only includes boundaries associated with Middle Eastern and North African states; in particular, not all the boundaries in the Mediterranean Sea are included. A number of potential boundaries associated with micro-territories such as the Spanish enclaves in Morocco, and the federal boundaries of the United Arab Emirates have been excluded. Their inclusion would add approximately 20 more boundaries (Drysdale and Blake 1985; 124). To date, less than one fifth of the boundaries in **Table 7.1** have been formally agreed; worldwide over a quarter have been agreed, so that the Middle East and North Africa lag somewhat behind the rest of the world in maritime boundary delimitation, (Smith, 1985).

Coastal states are entitled to claim several different types of offshore jurisdiction, all of which may involve international boundary delimitation. The main types of jurisdiction are territorial waters, the continental shelf, and the exclusive economic zone. Principles governing their functions and delimitation are set out in the 1982 U.N. Convention on the Law of the Sea. Territorial waters are now standardised to 12 nautical miles offshore. Here, the coastal state exercises a high degree of sovereignty, though ships of other states have the right of innocent passage.

TABLE 7.1
MARITIME BOUNDARIES IN THE MIDDLE EAST & NORTH AFRICA

	Agreed	Not Agreed	Total
Mediterranean Sea	3	20	23
Black Sea	1	1	2
The Gulf	7	7	14
Indian Ocean	0	5	5
Red Sea	1	14	15
Atlantic Ocean	0	3	3
Total	12	50	62

TABLE 7.2 : BOUNDARY STATUS IN 1986

	Agreement	Active Dispute	Potential Dispute	No Problem
Mediterranean Sea	3	3	5	12
Black Sea	1	0	0	1
The Gulf	7	6	0	1
Indian Ocean	0	0	1	4
Red Sea	1	2	6	6
Atlantic Ocean	0	1	0	2
Total	12	12	12	26

See Appendix A

Continental shelf rights give the coastal state exclusive control of the seabed resources of their own continental shelves, notably oil and natural gas. All coastal states are also entitled to a 200 nautical mile (1 nautical mile = 1.156 statute miles) exclusive economic zone (EEZ) under the 1982 U.N. Law of the Sea Convention, in which all resources of the seabed and waterbody are the property of the coastal state. As far as the Middle East and North Africa are concerned, all the boundary agreements shown in **Table 7.1** and in **Appendix A** are continental shelf agreements. Only one formal territorial water agreement has been traced, between the United Kingdom and Cyprus in 1961, outlining territorial water boundaries for the two U.K. sovereign bases in Cyprus. Although three Middle East and North African states have apparently declared EEZ's (Morocco 1981, Oman 1981, and P.D.R. Yemen 1978) none has delimited any E.E.Z. boundaries with neighbours (U.S. Department of State 1985). Morocco, Oman and P.D.R. Yemen all enjoy extensive and productive fisheries offshore, so that an EEZ is clearly advantageous. Exclusive economic zones also give the coastal state certain other rights such as control of scientific research, and preservation of the marine environment and it is surprising that states of the region have been so slow to declare EEZs. When they do, it can be generally assumed that a single maritime boundary will serve all purposes.

The following sections consider briefly the characteristics of selected boundary agreements, and major boundary problems. A large number of potential maritime boundaries in the Middle East and North Africa present no major political difficulty, though there may be tricky legal and technical questions to be answered **(see Table 7.2)**. Several may also be dependent upon the solution to disputes in the same vicinity before delimitation is possible.

7.2 Boundary Agreements

The earliest maritime boundary agreements were in the Persian/Arabian Gulf where large offshore oil and gas deposits lay in relatively shallow waters shared by eight riparian states. The first was the Bahrain-Saudi Arabia agreement of 1958. A median line was the basis, but an unusual method of delimitation was used, bisecting lines joining predetermined landmarks on either side of the sea. Small offshore islands were ignored. In the Fasht Bu Safaa hexagon the seabed was agreed to belong to Saudi Arabia, but the parties agreed to share the revenues from oil. This appears to be the earliest example of a shared economic zone as a temporary or permanent solution to problems of sovereignty. This, or similar arrangements illustrated in **Figure 7.1,** might usefully be applied elsewhere in the region.

The Iran-Saudi Arabia agreement of 1969 had to overcome several difficulties, including two disputed islands in the centre of the Gulf, and the presence of several smaller islands off the coasts. The agreement gave Farsi island to Iran and Arabi island to Saudi Arabia. Each island was allocated a twelve nautical mile territorial sea, round which the median line boundary was to deviate. Small islands were ignored, and the parties agreed not to drill for oil within 500 metres of the boundary line (El-Hakim 1979; 91-95).

Only one agreement has been concluded in the Red Sea, between Sudan and Saudi Arabia in 1974. The chief incentive was the discovery of metal-bearing muds in the deepest parts of the seabed in 1966, which were economically attractive. Although the most favourable deeps lie on the Sudanese side of the theoretical median line, it was agreed that both states would have equal rights to exploit resources at depths below 1,000 metres. A Saudi Arabian-Sudanese Joint Red Sea Commission was set up using Saudi capital to explore the possibility of commercial mining of zinc, copper, silver, gold, lead, and iron. Commercial exploitation should begin in 1988, with revenues shared between the parties (Mustafa 1984; 7-17).

Mediterranean boundary delimitation has been slow for a number of reasons. The prospects for oil and gas discoveries are not very good because of the prevalence of deep water. The presence of many islands and physically complex coasts has deterred boundary making, while the frequency of political disputes between 18 coastal states has been a further obstacle. Italy has led the way, concluding agreements with Yugoslavia (1970), Greece (1980), Tunisia (1971) and Spain (1978). The only other formal agreement in the Mediterranean is between France and Monaco in (1984). Tunisia is the only Middle Eastern or North African state to have reached a boundary agreement in the Mediterranean Sea. With the judgments of the International Court of Justice in cases between Libya and Tunisia (1982) and Libya and Malta (1985) these states can be expected to conclude formal boundary agreements in the near future.

The Tunisia-Italy agreement of 1971 (in force from 1978) appears to have favoured Tunisia. A median line boundary was fixed which disregarded the Italian islands of Pantelleria, Linosa, Lampedusa and Lampione. The islands were given enclaved continental shelf areas extending to 12 or 13 nautical miles. The resultant boundary could create difficulties when the terminus of the boundary between Libya and Tunisia is determined. The Italian islands were discounted in favour of Tunisia, but Italy may not wish to discount them in favour of Libya or Malta (Prescott 1985; 301).

The two cases which the International Court of Justice was asked to decide upton (Libya-Tunisia and Libya-Malta) illustrate the extraordinary complexity of international boundary making. Libya and Tunisia asked the Court to take account of all circumstances relevant to producing an equitable result. The Court judgement was given in 1982 after four years of hearings, and the accumulation of many volumes of pleadings. Libya argued for a boundary running roughly due north following the natural prolongation of the landmass northwards based on geology. Tunisia claimed a boundary roughly at 45° which would reflect the direction of the Tunisian coast, and the natural prolongation of Tunisia's landmass eastwards as shown by bathymetry and geomorphology. The Court rejected geological arguments, and emphasised the special geographical circumstances of the change in direction of Tunisia's coast, and the islands of Djerba and Kerkennah. It did not take into account the relevant wealth of the parties which had featured in the Tunisian pleadings. The recommended boundary consisted of a sector from the land boundary at 26°, reflecting the spatial conduct of the parties at sea over about 80 years (including oil concessions). A second sector, beginning at the latitude where the Gulf of Gabes reaches its most westerly point, would give half effect to the Kerkhennah islands, resulting in a line at 52°, **(see Figure 7.2).** Much has been written on this controversial case, but perhaps its most significant feature is the application of several methods and principles to the same delimitation. The Court's proposal gives the parties a share of the seabed approximately in proportion to the lengths of their coastline opposite the sea in question, which may be considered some indication of success.

The International Court judgment on the Libya-Malta continental shelf dispute was received in 1985. Malta had argued strongly for an equidistance line, while Libya depended on geological arguments showing the natural prolongation of the African land mass to a rift zone in the seabed approximately two thirds of the way to Malta. The Court took the view that the 1982 U.N. Convention on the Law of the Sea enabled states to claim continental shelf up to 200 miles offshore whatever the geological circumstances, and that there was no reason to ascribe any role to geological or geophysical factors. It also argued that the equidistance method had to be demonstrated to lead to an equitable result, and noted a disparity in the length of the relevant coasts of Malta (24 miles) and Libya (192 miles). This warranted the adjustment of the median line 18 nautical miles northwards. Arguments about the relative economic position of the two states were rejected (The Times, June 5 1985). Only a short length of boundary was fixed by the Libya-Malta judgment since Italy has substantial interests to the east of the line. Here, three parties will have to negotiate a very complex allocation at some time in the future.

7.3 Disputes

The most serious disputes and potential disputes in the seas of the Middle East and North Africa are in the Mediterranean and The Gulf. To some extent the relative absence of disputes in the Red Sea may be the result of the riparian states not having raised the question of maritime boundaries. When they do, more problems could arise. There are already disputes between Israel and Egypt affecting about 700 metres along the Gulf of Aqaba in the Taba area. At the southern end of the Gulf of Aqaba the islands of Sanafir and Tiran are claimed by Egypt and Saudi Arabia (El Hakim 1979; 137). The Sudan-Egypt boundary could also give rise to difficulties. While the Egypt-Sudan boundary seems clearly to be the 22nd parallel, Sudan could argue that the administrative boundary of 1902 should be the starting point for offshore delimitation, which would greatly favour Sudan (Brownlie 1979; 111-120).

Political and territorial questions seem likely to impede Mediterranean boundary delimitations very seriously. The most dangerous and the best documented of these is the Aegean Sea dispute between Greece and Turkey. There are also a number of micro-territorial disputes which could have a fundamental effect on the Mediterranean boundary system and these are too often overlooked, **(see Table 7.3).** The most difficult case concerns the five "plazas de soberania" owned by Spain on or near the coast of Morocco.

Since most of the islands are capable of sustaining human habitation, Spain's theoretical offshore claims could be extensive. The process of boundary delimitation would be complex, but it is unlikely to proceed because Morocco claims the territories. However, Spain has held the cities of Ceuta and Melilla for centuries and is unlikely to cede them to Morocco. The problem of the "plazas de soberania" is politically linked to the question of Gibraltar, which Spain claims. Britain is likely to come to some arrangement with Spain for the eventual return of Gibraltar to Spanish control, but the process could take many years. Meanwhile the Spanish view is that under the Treaty of Utrecht (1713) by which Britain acquired Gibraltar, offshore sovereignty was never conceded by Spain. The British do not agree, and there is periodic tension along the median line between Spain and Gibraltar in Algeciras Bay. An offshore boundary eastwards could give Gibraltar a significant wedge of seabed in a strategically sensitive part of the Mediterranean.

The large area of seabed around Cyprus could possibly be sub-divided between three parties. The British Sovereign Bases were permanently ceded to Britain as part of the independence agreement for Cyprus in 1960. Akrotiri in the west, has a generous slice of territorial waters, which, if extended as a

TABLE 7.3

MICRO-TERRITORIAL DISPUTES IN THE MEDITERRANEAN

Territory	Land Area (Sq.Kms)	Contenders
Cueta	7.50	Spain/Morocco
Penon de Velez de la Gomera	0.20	Spain/Morocco
Penon de Alhucemas	0.20	Spain/Morocco
Islas Chafarinas	2.50	Spain/Morocco
Melilla	4.75	Spain/Morocco
Gibraltar	6.50	Spain/Britain
Northern Cyprus	3,423.00	Cyprus/Turkey
UK Sovereign Base - Dhekelia # UK Sovereign Base - Akrotiri #	256.00	Britain/Cyprus
Gaza Strip	362.00	Israel/Palestine

continental shelf or EEZ boundary could reach the boundary with Egypt. Britain is not likely to press claims on behalf of the Sovereign Bases, but would probably have to declare an interest if serious boundary negotiations between Cyprus and Egypt were to begin. Potential Cypriot offshore boundaries were complicated by the Turkish invasion of Cyprus in 1974 and the **de facto** partitioning of the island. Northern Cyprus declared itself to be independent (U.D.I.) in 1983, and it might be tempted to claim offshore waters because of known deposits of minerals, though not of hydrocarbons (Bastianelli 1983; 2). The Turkish Republic of Cyprus has only been recognised by Turkey however, and its offshore claims would not be regarded as internationally valid.

The Gaza Strip is a heavily populated region of former Palestine, with a coastline of some 45 Km. Occupied by Egypt from 1948 until 1967, and taken by Israel in 1967, its political future is uncertain but it could become part of some future Palestinian political entity, based primarily on the landlocked West Bank. An offshore area could be claimed, which might be highly symbolic for the Palestinians, and could contain useful resources. Similarly, if Lebanon ever became permanently partitioned the allocation of offshore areas to the micro-states would probably become an important issue.

The majority of maritime boundary disputes in the Gulf are associated with the ownership of islands. Until agreement is reached over the sovereignty of these islands, the boundary system is unlikely to advance much beyond its present state. The long-running dispute between Kuwait and Iraq over Bubiyan and Warbah islands causes most concern because Iraq is preoccupied with their strategic value as well as the effect their tenancy would have upon seabed allocations in the northern Gulf. Ever since Iraqi forces crossed into Kuwait in 1972 to support their territorial claims, there has been fear of Iraqi intentions in Kuwait (Amin 1984; 43). No boundary can be drawn between Qatar and Bahrain because Bahrain claims the Huwar islands, which lie close to Qatar, on historic grounds. Qatar claims the tiny Bahraini islet of Fasht al Dibel, the scene of an incident in April 1986 (The Times, April 28 1986).

In 1971 Iran occupied Abu Musa island and the Tumb islands at the entrance to the Gulf, thus sparking a potentially explosive dispute with the United Arab Emirates, (Amin 1981, 163). Iran's claims to Abu Musa are based on Persian occupation until the nineteenth century, and the need to guard the entrance to the Gulf in the interests of national security. Abu Musa is claimed by Sharjah, and the Tumbs by the ruler of Ras al-Khaymah, but both remain firmly in Iranian control.

A third example of disputed islands in the Gulf are Qaru and Umm al-Maradim, located in the former Neutral Zone between Kuwait and Saudi Arabia. The 1969 Kuwait-Saudi Arabian agreement to partition the Neutral Zone did not decide ownership of the small islands. Saudi Arabia maintains that they are subject to co-sovereignty status, while Kuwait claims full sovereignty. The dispute is impeding formal boundary agreements between Kuwait and Saudi Arabia, and with Iran.

7.4 Conclusion

The 1982 Law of the Sea Convention boosted national interest in maritime boundary delimitation, and many relatively straightforward agreements are likely to be concluded over the next few years. The most difficult areas are the semi-enclosed seas surrounded by several riparian states such as the Gulf, the Red Sea, and the Mediterranean Sea involving opposite and adjacent boundaries. Boundary agreements are desirable chiefly to facilitate resource exploitation and environmental management, while the absence of agreement clearly increases the risk of international conflict. In the Mediterranean, history and geography make boundary delimitation a long and sometimes costly process. In the Gulf, uncertainties over island sovereignty, and the traditional Iranian-Arab rivalry will make further progress slow. Several boundaries could be agreed in the Red Sea, though a number of potential legal differences remain (Marston, 1984; 47-62). It may be suggested that more regional effort could be made to solve the outstanding problems, in much the same way that Mediterranean, Gulf, and Red Sea states have agreed to cooperate as part of the United Nations Environmental Programme Regional Seas Programmes to combat pollution. A number of compromise arrangements have been devised and implemented in various parts of the world to facilitate resource exploitation without prejudicing the question of national sovereignty over islands or the seabed. If this, or some other means is not found soon, the Middle East and North Africa could be left far behind the rest of the world in establishing an agreed regime for the proper management of ocean space.

References

Amin, S A (1981), **International and Legal Problems of the Gulf,** Menas Press Ltd, Wisbech, Cambridgeshire

Amin, S A (1984), **Political and Strategic Issues in the Persian-Arabian Gulf,** Royston, Glasgow

Bastianelli, F (1983), "Boundary delimitation in the Mediterranean Sea". **Marine Policy Reports** Vol. 5 no. 4, University of Delaware, College of Marine Studies

Blake, G H (1982), **Maritime Aspects of Arabian Geopolitics.** Arab Papers No. 11. Arab Research Centre, London

Brownlie, I (1979), **African Boundaries, a Legal and Diplomatic Encyclopaedia**, Hurst, London for the R.I.I.A.

Drysdale, A and Blake G H (1985), **The Middle East and North Africa : a Political Geography**, Oxford U.P., New York

El-Hakim, A A (1979), **The Middle Eastern States and the Law of the Sea**, Manchester U.P., Manchester

Marston, G (1984), "Potential legal problems" in A M Farid (ed) **The Red Sea,** Croom Helm, London pp 47-62

Mustafa, Z (1984), "Red Sea resources" in A M Farid (ed) **The Red Sea**, Croom Helm, London pp 7-17

Prescott, J R V (1985), **The Maritime Political Boundaries of the World,** Methuen, London

Smith, R W (1982), "A geographical primer to maritime boundary-making". **Ocean Development and International Law Journal.** 12(1/2), 1-22

Smith, R W (1986), **Exclusive Economic Zone Claims, An Analysis and Primary Documents**, Martinus Nijhoff, Dordrecht

U.S. Department of State (1985). Limits in the Seas No. 36 **National Claims to Maritime Jurisdictions.** Washington

APPENDIX A : MARITIME BOUNDARY STATUS OF MIDDLE EAST AND NORTH AFRICAN STATES IN 1986

	Boundary Agreed	Boundary not Agreed: Active Dispute	Boundary not Agreed: Potential Dispute	Boundary not Agreed: No known Problems
MEDITERRANEAN				
Algeria-Spain	-			*
Algeria-Morocco	-			*
Algeria-Italy	-			*
Algeria-Tunisia	-			*
Cyprus-Turkey	-	*		
Cyprus-Syria	-		*	
Cyprus-Lebanon	-		*	
Cyprus-Israel	-			*
Cyprus-Egypt	-			*
Cyprus-Greece	-			*
Egypt-Libya	-		*	
Egypt-Israel	-			*
Egypt-Greece	-		*	
Israel-Lebanon	-			*
Lebanon-Syria	-			*
Libya-Greece	-			*
Libya-Malta	1985-ICJ			
Morocco-Spain	-	*		
Syria-Turkey	-		*	
Tunisia-Italy	1978			
Tunisia-Malta	-			*
Tunisia-Libya	1982-ICJ			
Turkey-Greece	-	*		
BLACK SEA				
Turkey-USSR	1981			
Turkey-Bulgaria				*
THE GULF				
Bahrain-S.Arabia	1958			
Bahrain-Iran	1972			
Bahrain-Qatar	-	*		
Iran-Iraq	-	*		
Iran-Kuwait	-	*		
Iran-S.Arabia	1968			
Iran-Qatar	1969			
Iran-UAE	1974(Dubai)			
Iran-UAE	- Other Sectors			
Iraq-Kuwait	-			
Kuwait-S.Arabia	-			
Oman-UAE	-			

Appendix A - continued

	Boundary Agreed	Boundary not Agreed		
		Active Dispute	Potential Dispute	No known Problems
Oman-Iran	1974			
Qatar-UAE	1969			
Qatar-S.Arabia	-			
INDIAN OCEAN				
Oman-UAE (N)	-			*
Oman-UAE (S)	-			*
Oman-Iran	-			*
S. Yemen-Oman	-		*	
S.Yemen-Somalia	-		*	
RED SEA				
Djibouti-S.Yemen	-		*	
Djibouti-Ethiopia	-		*	
Djibouti-N.Yemen	-		*	
Egypt-Israel	-	*		
Egypt-S.Arabia	-	*		
Egypt-Sudan	-		*	
Egypt-Jordan	-			*
Ethiopia-Sudan	-			*
Ethiopia-S.Arabia	-			*
Ethiopia-N.Yemen	-			*
Israel-Jordan	-		*	
Jordan-S.Arabia	-			*
S.Yemen-N.Yemen	-		*	
S.Arabia-Sudan	1974			
S.Arabia-N.Yemen	-			*
ATLANTIC				
Morocco-W.Sahara	-	*		
Morocco-Spain (Canaries)	-			*
Morocco-Portugal (Madeira)	-			*

Sources :

SP Jagota (1985), **Maritime Boundary**, Martinus Nijhoff.
JRV Prescott (1985), **The Maritime Political Boundaries of the World**, Methuen, London.
US Department of State, **Limits in the Seas**, (series)

Fig. 1 Maritime Boundaries and Economic Zones

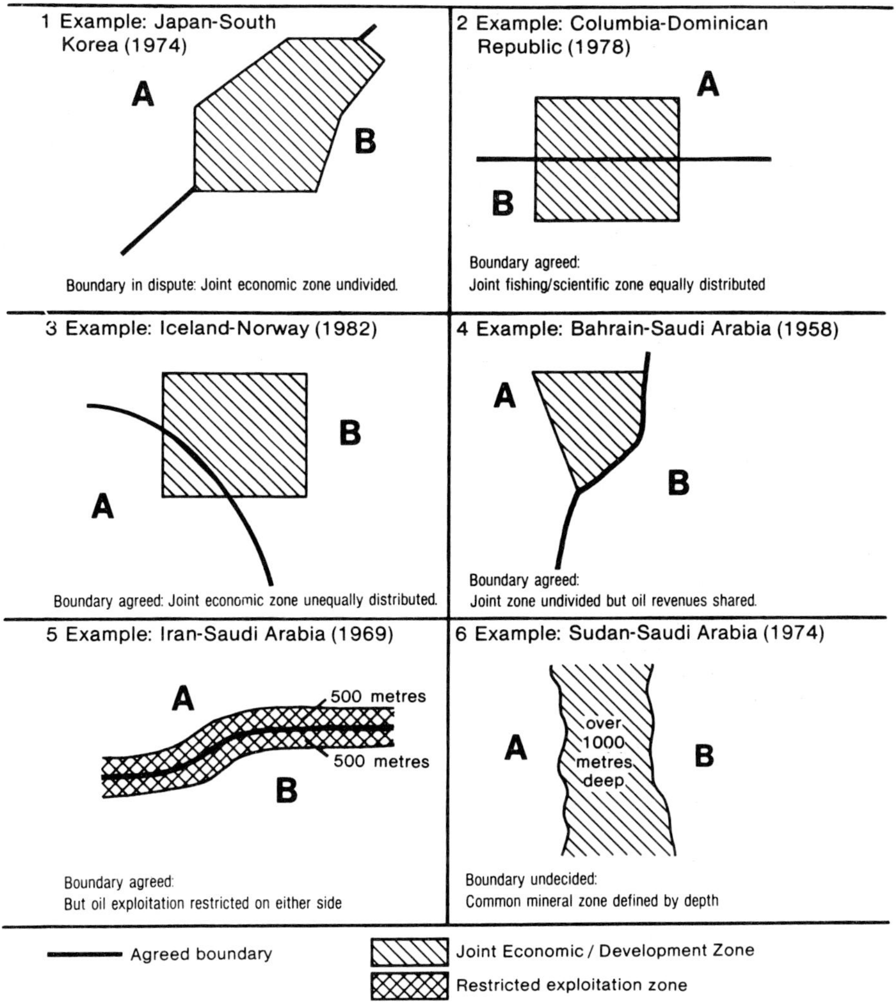

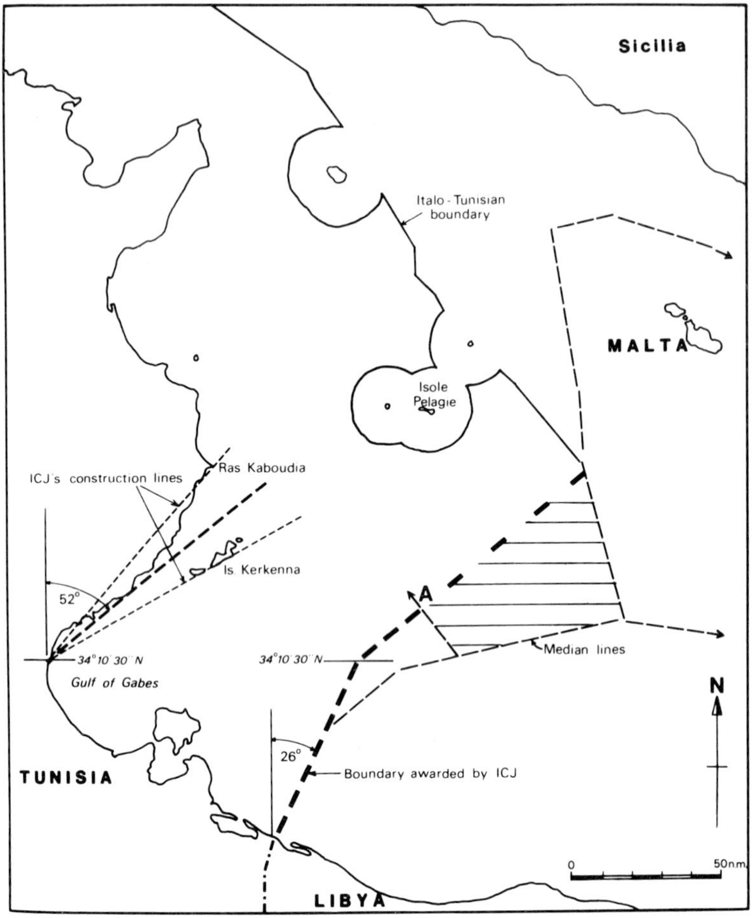

FIG 2. Libya-Tunisia Maritime Boundary as recommended by the International Court of Justice

8 INTERNAL WATERS AS EXTENSIONS OF STATE SOVEREIGNTY IN THE MEDITERRANEAN SEA : THE CLAIMS OF NORTH AFRICAN STATES TO STRAIGHT BASELINES & HISTORIC BAYS

Stephen Langford

8.1 Introduction

The most significant difference between maritime space and land territory is that, with the exception of internal waters, offshore jurisdiction is not absolute, i.e. the coastal State exercises something less than full sovereignty. Complete territorial sovereignty stops, under usual circumstances, at the low-water line along the coast, which is defined in the 1982 Convention on the Law of the Sea as the "normal baseline" for the measurement of the territorial sea (Article 5). However, Article 8 (1) of the 1982 Convention refers to "waters on the landward side of the baseline of the territorial sea" as forming "part of the internal waters of the State". This is reference to the fact that, under particular geographical circumstances, international law makes provision for baselines to be drawn across water between salient points on land, rather than following the low-water line along the coast in all its sinuosities. Thus, rather than a single baseline - the low-water line - defining the place along the coast where land dominion gives way to maritime jurisdiction, provision is made for baselines departing from the coastal configuration, to define the external limit of internal waters.

Internal waters, (or sometimes "inland", or "national" waters), are those over which a State exercises complete sovereignty; in other words, the water area is assimilated to land for sovereignty purposes. Such waters are created most often where closing lines are drawn across the mouths of rivers which flow directly into the sea (Article 8 of the 1982 Convention), or across the mouths of bays which conform to the juridical definition of bays set down in Article 10 of the 1982 Convention. In addition, internal waters are created where a coastal State implements a system of so-called "straight baselines" or lays claim to an "historic bay". This paper is concerned with straight baseline and historic bay claims by North African States with coasts on the Mediterranean Sea; internal water claims, which extend State sovereignty, can be viewed in contravention of the international law rules governing the use of such baselines.

8.2 Straight baselines

Baselines are important, not only because they define the outer limits of a State's internal waters - over which the State exercises complete sovereignty - but also because all the zones of national jurisdiction offshore (e.g. continental shelf, exclusive economic zone) are measured from the baselines of the territorial sea. Thus, the further the baselines of the territorial sea depart from the coastline itself, the further a coastal State's offshore jurisdiction, of whatever type, is pushed seaward.

As regards baselines, the 1982 Convention repeats the provisions of the 1958 Convention on the Territorial Sea and Contiguous Zone, in allowing for the normal baseline - the low-water mark on the coast of mainland and islands - to be derogated from in specific geographical circumstances. A system of straight baselines may be employed:

(i) in localities where the coastline is deeply indented and cut into;

(ii) where there is a fringe of islands along the coast in its immediate vicinity;

(iii) where because of the presence of a delta and other natural conditions the coastline is highly unstable.

However, under the 1982 Convention, the employment of these straight baselines is dependent upon the fulfilment of various criteria:

(a) that the drawing of straight baselines must not depart to any appreciable extent from the general direction of the coast;

(b) that the sea areas lying within the lines must be sufficiently closely linked to the land domain to be subject to the regime of internal waters;

(c) that the system of straight baselines may not be applied in such a manner as to cut off the territorial sea of another State from the high seas or an exclusive economic zone.

Further, where the coast is deeply indented and cut into,

or fringed by islands in its immediate vicinity, account may be taken in determining particular baselines, of economic interests peculiar to the region concerned, the reality and importance of which are clearly evidenced by long usage.

It appears that these provisions, which originated in the International Court of Justice judgement in the 1951 Fisheries Case between the United Kingdom and Norway (I.C.J. Repts. 1951), were intended for use in coastal geographic circumstances resembling those present on the northern coast of Norway. They were not intended for widespread application. However, their verbatim inclusion in both the 1958 Geneva Convention on the Territorial Sea and Contiguous Zone (Article 4), and now the 1982 Convention (Article 7), appears to have encouraged States to interpret the rules without reference to the original intent. This has been exacerbated by the fact that these provisions are open to a wide variety of subjective interpretations.

The importance of recognising straight baselines which may be regarded as excessive, or misapplications, of the rules governing their use, is twofold. First, they increase the area of internal waters under the exclusive sovereignty of the State. Although innocent passage, and hence freedom of navigation rights, are safeguarded by Article 8 (2) of the 1982 Convention, straight baselines may enclose considerable expanses of water. Secondly, as the baseline moves further and further away from the shoreline, so too do the other zones of national jurisdiction. In a narrow semi-enclosed sea like the Mediterranean, this may be especially important for drawing boundaries between States; particularly so for States in a relationship of oppositeness, as the equidistance line, for example, is so dependent on baselines for its construction.

Twelve Mediterranean States have legislation claiming straight baselines along all, or parts of their coasts; namely: Albania, Yugoslavia, Spain, Egypt, France, Syria, Turkey, Tunisia, Malta, Italy, Morocco and Algeria. Of these, those of Tunisia, Morocco and Algeria, will be the subject of scrutiny here.

8.3 Straight baselines in the southern Mediterranean Sea

Tunisia's straight baselines, **(see Fig. 8.1)**, include the surrounding of the Kerkennah Islands. This is held to be a legitimate use of fringing islands (Prescott 1985; 297). De Guttry disagrees; she holds that the straight baselines around these islands do not follow "the general direction of the coast" (De Guttry 1984; 397), presumably because they also embrace the drying reefs seaward of these islands. However, Tunisia legitimises her claims by invoking the legal doctrine of "historic waters" (Gioia 1984; 327-376). The 1982 Convention makes no provisions concerning historic waters, although it does mention "historic bays" as exceptions to the rules governing bay closing lines (Article 10 (6)). The claim of historic waters may be legitimate, but it does not appear to be a valid reason for employing straight baselines that depart from "the general direction of the coast", the primary criterion laid down in the 1951 Fisheries judgement for validating the use of a straight baseline system. It was, therefore, a pity that in the Tunisia-Libya Continental Shelf Case, the legitimacy of these Tunisian baselines was not pronounced upon (I.C.J. Repts. 1982).

Morocco proclaimed straight baselines by a Decree of 1975. This clearly places the five Spanish "plazas de soberania" on or near the coast of Morocco, (which consist of two coastal enclaves - Ceuta and Melilla - and three sets of islands), behind the Moroccan straight baselines, thereby denying them any maritime zones to which they may be entitled. Indeed, the Chafarinas Islands are used as basepoints in the Moroccan straight baseline system. This system appears to directly contradict Article 7 (6) of the 1982 Convention, which states that: "The system of straight baselines may not be applied by a State in such a manner as to cut off the territorial sea of another State from the high seas or an exclusive economic zone". The justification for Morocco's action seems to be paragraph 10 of the Organisation for African Unity's 1972 "Declaration on Issues of the Law of the Sea", which concerned the non-recognition of exclusive economic zone regimes in the case of territories under colonial domination (Symmons 1979; 57). At the Caracas session of the Third United Nations Conference on the Law of the Sea, Morocco stated it also included non-recognition of the territorial sea in cases of territories under colonial domination. This was seemingly a direct reference to its viewpoint on the Spanish enclaves and islets. If Morocco were able to secure sovereignty over the Spanish islets off her coast, it would appear perfectly appropriate for her to draw straight baselines claiming the islands as a fringe in its immediate coastal vicinity. However, at present, she does so notwithstanding the fact that the islets are under Spanish sovereignty. The baselines drawn along other parts of the coast, though apparently not departing to any appreciable extent from the general direction of the coast, do not appear to be justified under any rule of law, as the coastline is not deeply indented and cut into, but rather is generally smooth.

The same may be said for the Algerian straight baselines proclaimed in a Decree of 1984. These straight baselines extend along the entire length of the coast of Algeria. Several bays have been enclosed, and tiny islets, such as Rachgoun and Habibas, have been used as basepoints, even though they are isolated and cannot be said to constitute a "fringe" of islands. In places, the baselines appear to depart from the coast, though from the nautical chart studies, it was not apparent what were being used as basepoints.

Egypt has also made a claim to straight baselines in the Mediterranean, although its Mediterranean coast is not appropriate for straight baseline drawing. It does not appear to have produced a chart showing these straight baselines (Prescott 1985; 298), perhaps for that very reason.

Thus, of the five North African States on the Mediterranean Sea, only Libya has no straight baseline legislation. Considerable doubt is seen to surround the legitimacy of each of the other four States' claims. Straight baselines may prove to be sources of dispute in the delimitation of maritime boundaries between neighbouring States, where a State objects to the use by its neighbour of such a method of baseline construction. This is likely to happen where either the baseline's orientation, or its distance from the coast, is such as to be perceived as disadvangageous to a neighbouring opposite or adjacent State, in the drawing of their common boundary line. Indeed, Libya objected to Tunisia's straight baselines around the Kerkennah Islands in their continental shelf dispute because of their potential disadvantageous effect on the boundary to be drawn (Gioia 1984; 327-376). Other situations like these are quite possible in the Mediterranean. Disputes over even small areas of seabed or seaspace may provoke major disputes, due to the restrictions placed on extending offshore jurisdiction by the proximity of neighbouring States, exacerbated by uneasy inter-State relations. In both the Tunisia-Libya (I.C.J. Repts. 1982) and Malta-Libya Continental Shelf Cases (I.C.J. Repts. 1985) the International Court of Justice refrained from adjudicating on disputed baselines. Instead, it was able to use delimitation techniques which avoided making use of disputed basepoints.

8.4 "Historic bays"

The 1982 Convention sets out, in Article 10, particular rules concerning the drawing of straight lines across bays. In general terms, the closing line of the bay must not exceed 24 nautical miles, and the area of waters enclosed behind such a line must be as large, or larger than, the area of a semi-circle whose diameter is the closing line of the bay. Article 10 (b), however, reads in part: "The foregoing provisions do not apply to so-called 'historic bays' ...", which are never defined in the Convention.

The problem of "historic bay" claims has taxed the minds of lawyers for a considerable length of time, but largely remains unsolved. Such claims have been invoked by States as a justification for closing "bays", which meet neither the criteria for the length or permissible closing line, nor the semi-circle test, desribed above. Attempts to codify a definition of a "historic bay" have been unsuccessful, although many authors have themselves provided definitions. Bouchez provides the following definition of a historic bay:

> "Historic ... (bays) are waters over which the coastal State, contrary to the generally applicable rules of international law, clearly, effectively, continuously, and over a substantial period of time, exercises sovereign rights with the acquiescence of the community of States". (Bouchez 1964; 281)

Thus, a historic bay must be claimed to be under the sovereignty of the coastal State; the claim must be confirmed by the exercise of sovereignty for a sufficiently long period; and it must be acquiesced in by other interested States.

Historic bay closing lines form part of a coastal State's baselines. The waters enclosed by the baseline are "internal waters", and thus assimilated to land from the point of view of territorial sovereignty. Moreover, no right of innocent passage exists through internal waters created by historic bay claims. The enclosure of large "bays" may significantly shift seaward the zones of national jurisdiction, based on their closing lines, and, as a consequence, affect the delimitation of boundaries with neighbouring States.

8.5 "Historic bays" along the southern Mediterranean coast

There are several claims to historic bays in the Mediterranean. These are made by four of its coastal States; namely Libya, Italy, Tunisia and Egypt. The claim of Italy to the Gulf of Taranto will not be considered here.

The most noteworthy historic bay claim in the Mediterranean is that of Libya to the Gulf of Sirte, made in 1973. This claim has been actively challenged by the United States on several occasions (Spinnato 1983; 65-85), and has twice led to military confrontations with the United States, in 1981 (Francioni 1980/81; 85-109, Adam 1982; 1025-1028) and 1986 (The Times 26 March 1986). Several states have protested the enclosure of the Gulf of Sirte, which converts international waters into internal waters, as an unlawful abuse of the freedom of the high seas. The most vociferous of these has been the United States, although protests have also been lodged, for example, by the European Community and Malta. The majority of States have consistently refused to acknowledge Libya's 1973 declaration of sovereignty over the Gulf's waters (Blum 1986; 674-676; Prescott 1985; 298). Only Burkina Faso (The Times 26 March 1986) and Syria (Blum 1986; 674) accept the claim to ownership of the Gulf, and apparently only one Western commentator on the subject upholds the

Libyan case (Rousseau 1974; 1179).

The closing line of the Gulf of Sirte is defined as the parallel of latitude 32° 30' north. This measures 296 nautical miles, and at one point on the closing line the nearest coast is 96 nautical miles away. This clearly has a considerable effect on Libya's territorial sea claim, which extends to 12 nautical miles from the Gulf of Sirte baseline. It promises to be an even greater problem in the respective future delimitations of continental shelf and/or exclusive economic zone boundaries with Malta and Greece. Neither Malta nor Greece is likely to recognise any Libyan boundary claim which makes use of a baseline that, in effect, puts the opposite coast of Libya up to 96 nautical miles nearer to their own coasts. Indeed, a Greek protest at the enclosure of the Gulf of Sirte probably reflects this concern. Use of the "historic bay" concept is the only "legitimate" means for Libya to draw a closing line across the Gulf, as the bay does not meet the semi-circle test and its closing line is far too long to comply with the rules defining a legal bay. Moreover, the area of internal waters enclosed is 70,560 square kilometres, or approximately 2.8% of the Mediterranean Sea.

The historic bay claim to the Gulf of Sirte has two bases. First, that Libya has exercised sovereignty over the waters throughout history and without any dispute; and secondly, that sovereignty over the bay is necessary for the national security of Libya. The overwhelming view of writers on the subject is that there is no evidence that Libya has exercised exclusive sovereignty over the claimed waters, even during the country's brief history, and that even if it had, there has clearly been no acquiesence in the claim by interested States. Thus, one is forced to conclude that the enclosure of the Gulf of Sirte on historic grounds is illegal. The security argument has yet to be properly tested and may prove to have some justification, though the case cannot be fully discussed here.

Tunisia has claimed both the Gulfs of Tunis and Gabes as historic bays, **(see Fig.8.1)**. The Gulf of Tunis has long been considered as a historic bay by many legal writers, but proof of the existence of such a claim seems unnecessary, as the bay fulfills the 1982 Convention's requirements for juridical bay status. It has a closing line of 23 nautical miles and meets the semi-circle test requirement. Neverthless, it can be shown that the historic bay claim is valid. Indeed, Libya, who vigorously contested the validity of Tunisia's baselines during the dispute over the delimitation of their respective adjacent continental shelves, appeared to accept the Tunisian claim to the Gulf of Tunis as a historic bay (Gioia 1984; 332-343).

For obvious geographical reasons, Libya concentrated her efforts on demonstrating the invalidity of Tunisia's baselines drawn in the region of the Gulf of Gabes, as their acceptance would have a considerable effect in potentially diverting the continental shelf boundary to the disadvantage of Libya. Gioia is able to show that although several legal writers have cited the Gulf of Gabes as a possible example of an historic bay, their opinion has been based on that of Gidel, who, whilst he did not hesitate in describing the Gulf of Tunis as a historic bay, was more cautious in his assertions concerning the Gulf of Gabes (Gioia 1984; 343-351).

The Gulf of Gabes was not explicitly claimed as a historic bay by Tunisia until 1973, when a closing line of 45 nautical miles was drawn. No historical justification was provided at the time of the claim, though such a justification was invoked during the dispute with Libya. This referred to historic rights, of great antiquity, over sedentary fisheries, which Tunisia held still to be applicable. Gioia argues that these historic rights were not to be regarded as indicative of the full sovereignty over the waters and seabed necessary for the waters to be regarded as "internal", before the claim of 1973 to historic bay status: indeed, when such rights had been held by Tunisia as indicative of her sovereignty, Gioia is able to provide examples of dissension by other States (Gioia 1984; 345-349). No claim was made for an historic bay before 1973; the exercise of jurisdiction over the waters claimed was not sufficient to be regarded as territorial sovereignty; and the claim was not acquiesced in: thus, the Gulf of Gabes was not in 1973 an historic bay. Since 1973, the Libyan protests against the enclosure of the Gulf of Gabes similarly deny the Gulf historic bay status. Neither Malta nor Italy have acquiesced in this Tunisian claim (Gioia 1984; 349-40).

Egypt has claimed all five of its Mediterranean bays as "inland waters" since 1951. The Bay of El Arab had been claimed as a historic bay in 1926. The only justification given for this treatment of the Bay's waters was "its geographical configuration" (Bouchez 1964; 221). The Bay of El Arab resembles a gentle curvature of the shoreline, enclosed by a straight line approximately 95 miles wide (Nixon 1981; 15). It clearly does not fulfil the semi-circle test, though such a test did not exist at the time the claim was made.

The Egyptian Royal Decree of 1951 did not mention the Bay of El Arab, but instead made inland waters "all the waters of the bays along the coast of the Kingdom of Egypt". This added the Bays of Solum, Abu Hashaifa, Pelusium and El Arish to Egypt's claimed internal waters. None of these bays satisfy the semi-circle test (Nixon 1981; 15). The Decree was the subject of a protest by Great Britain in 1951, in which it was claimed no historic bays existed along the coast of Egypt, and that those so claimed were mere curvatures of the coast. The United States also filed a protest to the Decree in the same year.

The Egyptian view was that proof of historic title or long

continuous usage was unnecessary for historic bay claims, and that a State might draw baselines in accordance with its own local conditions (Strohl 1963; 260-261). It is hard to see why Egypt felt it necessary to close these bays, though the most probable reason was security; it will be interesting to see in the future whether Egypt presses these claims when negotiating its boundaries with neighbouring States.

8.6 Conclusion

It is estimated that the area of internal waters claimed by all Mediterranean States, by virtue of straight baseline and historic bay claims, amounts to as much as 6.8 per cent of the Mediterranean Sea. With the notable exception of Libya, the area of the Mediterranean Sea covered by internal waters claimed by North African States is relatively small. However, the proportion of their potential areas of national jurisdiction - calculated on the basis of hypothetical equidistance lines - under internal waters gives more cause for concern. Percentage figures of 3.8% for Algeria, 3.0% for Egypt, 22.0% for Libya, 6.7% for Morocco, and 11.5% for Tunisia, as expressions of the internal water proportion of their respective potential areas of offshore jurisdiction, illustrate the extent to which these States have attempted to extend their full territorial sovereignty, in effect, by moving the coastline seaward. Clearly, such redefinitions of the coast as these have implications for the drawing not only of territorial sea boundaries, but also continental shelf and exclusive economic zone boundaries between neighbouring States, as they extend such zones of jurisdiction seaward. Bearing in mind the fact that the claims of the North African States both to straight baselines and historic bays are of doubtful validity, Fitzmaurice's words still hold true: "In short, a straight baseline system was, and is, simply a disguised method of claiming additional waters, or of altering in favour of the coastal State the status of water off its coasts" (Fitzmaurice 1954 p 421). This applies equally to the historic bay claims discussed.

8.7 Acknowledgement

Mr. Arthur Corner of the Cartographic Unit at Durham University kindly calculated areas of internal waters quoted in this chapter.

References

Adam, R (1981), 'L'incidente del golfo della Sirte'. **Revista di diritto internazionale.** Vol. 64, pp 1025-1028.

Blum, Y.Z. (1986), 'The Gulf of Sidra Incident'. **American Journal of International Law.** Vol. 80, pp 668-677.

Bouchez, L.J. (1964), **The Regime of Bays in International Law.** A.W. Sythoff, Leyden.

De Guttry, A. (1984), 'The delimitation of territorial waters in the Mediterranean Sea'. **Syracuse Journal of International Law and Commerce.** Vol. 11, pp 377-415.

Fitzmaurice, Sir G. (1954), 'The Law and Procedure of the International Court of Justice, 1951-54: Points of Substantive Law - 1'. **British Yearbook of International Law.** Vol. 31, pp 379-429.

Francioni, F. (1980/81), 'The Gulf of Sirte incident (United States v Libya) and international law'. **Italian Yearbook of International Law.** Vol. 5, pp 85-109.

Gioia, A. (1984), 'Tunisia's claims over adjacent seas and the doctrine of historic rights'. **Syracuse Journal of International Law and Commerce.** Vol. 11, pp 327-376.

International Court of Justice (1985), 'Case Concerning the Continental Shelf (Libyan Arab Jamahiriya/Malta) - Judgment of 3 June, 1985'. **Reports of Judgments, Advisory Opinions and Orders.** The Hague.

International Court of Justice (1982), 'Case Concerning the Continental Shelf (Tunisia/Libyan Arab Jamahiriya) - Judgment of 24 February, 1982'. **Reports of Judgments, Advisory Opinions and Orders.** The Hague.

International Court of Justice (1951), 'Fisheries Case - Judgment of 18 December 1951'. **Reports of Judgments, Advisory Opinions and Orders.** A.W. Sijthoff, Leyden.

Nixon D.W. (1981), 'A Comparative Analysis of Historic Bay Claims' in Libyan Reply (1980). **International Court of Justice - Continental Shelf (Tunisia/Libyan Arab Jamahiriya) - Technical Annexes.** Vol. 2, Annex II-3.

Prescott, J.R.V. (1985), **The Maritime Political Boundaries of the World.** Methuen, London.

Rousseau, C. (1974), 'Chronique des faits internationaux - Libye'. **Revue générale de droit international public.** Vol. 78, pp 1177-1179.

Spinnato, J.M. (1983), 'Historic and vital bays: an analysis of Libya's claim to the Gulf of Sidra'. **Ocean Development and International Law.** Vol. 13, pp 65-85.

Strohl, M.P. (1963), **The International Law of Bays.** Martinus Nijhoff, The Hague.

Symmons, C.R. (1979), **The Maritime Zones of Islands in International Law.** Martinus Nijhoff, The Hague.

United Nations (1983), **The Law of the Sea: Official text of the United Nations Convention on the Law of the Sea.** Croom Helm, London.

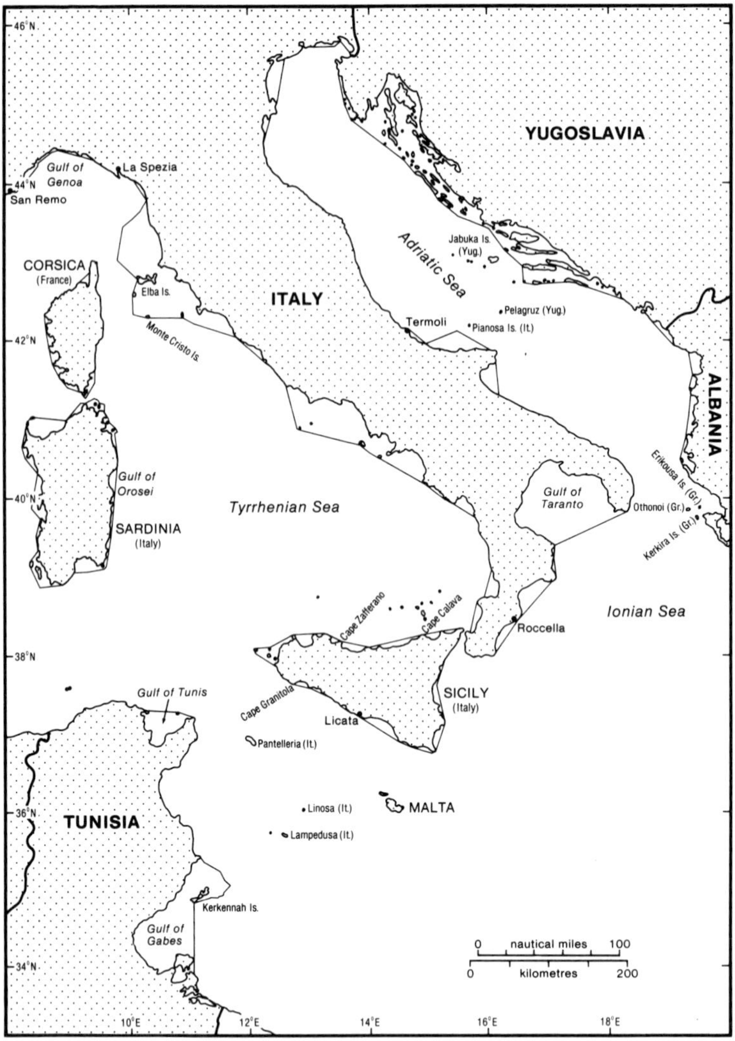

FIG 1. Tunisian straight baselines and historic bay claims in relation to the central Mediterranean region

ABOUT THE CONTRIBUTORS

Ewan Anderson is a Senior Lecturer in the Department of Geography at the University of Durham.

Gerald Blake is Reader in Geography at the University of Durham.

Carl Grundy-Warr is undertaking doctoral research in Geography at the University of Durham.

George Joffe is Consultant Editor for the Economist Intelligence Unit on the Middle East and North Africa and a Research Associate of the Centre of Near and Middle Eastern Studies, School of Oriental and African Studies, University of London.

Nurit Kliot is a Senior Lecturer in the Department of Geography at the University of Haifa, Israel.

Bahgat Korany is Director of the Arab Studies programme and Professor of Political Science at the University of Montreal, Canada.

Stephen Langford is undertaking doctoral research in Geography at the University of Durham.

Richard Schofield is a Senior Research Assistant in Geography at the University of Durham.

INDEX